The S

The
Smaller

THE HIDDEN WORLD

OF THE ANIMALS THAT

DOMINATE THE TROPICS

THE BELKNAP PRESS OF

HARVARD UNIVERSITY PRESS

Cambridge, Massachusetts
London, England

2005

PRINTED AND BOUND IN THE UNITED STATES OF AMERICA

Printed by DS Graphics of Lowell, Massachusetts, in Hexachrome® on 100-pound Parilux Silk
Bound by Acme Bookbinding of Charlestown, Massachusetts

All photographs by Piotr Naskrecki
Design by Tim Jones and Piotr Naskrecki
Composition by Tim Jones

Library of Congress Cataloging-in-Publication Data

Naskrecki, Piotr.
    The smaller majority / Piotr Naskrecki.
    p. cm.
    Includes bibliographical references and index (p.).
    ISBN 0-674-01915-6 (cloth : alk. paper)
    1. Invertebrates—Tropics. 2. Invertebrates—Tropics—Pictorial works.
    3. Animals—Tropics. 4. Animals—Tropics—Pictorial works. I. Title.

    QL109.N37 2005
    591.7'0913—dc22   2005046060

FIRST EDITION

This book is dedicated to my parents,

Helena and Wladyslaw,

for their endless love and support.

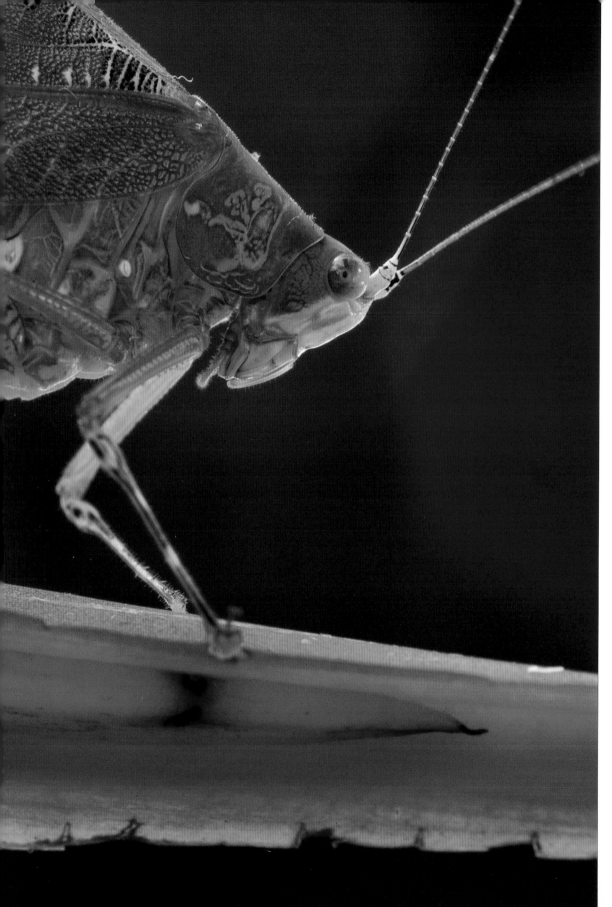

Leaf katydid *(Hyperphrona irregularis)* (Costa Rica)

Contents

Photograph by Wladyslaw Naskrecki

Prologue

How I Discovered the Smaller Majority

WE ARE BORN WITHOUT THE FEAR OF NATURE. Young children are fascinated with life around them, equally intrigued by a caterpillar or a dog. The fear of most creatures is instilled in us later in life by overly protective parents or teachers, peer pressure, and misguided media. By the age of ten most children either love or hate insects and other tiny organisms, and these feelings usually stay with them for the rest of their lives. I could never understand why small animals, including most amphibians and reptiles, evoke such polarized feelings. After all, how many people hate jaguars or elephants, things that can really hurt you? I think it can be explained in part by what psychologists call "prepared learning"—we are quick to learn the fear of snakes because millions of years of human evolution have favored individuals inclined to avoid them, even if most snakes are not venomous. But what about moths, spiders, or beetles? Why do most people find these animals repulsive, yet happily gorge on lobsters? When you think about it, their morphology is remarkably alike. Perhaps the negligible proportion of small animals that are indeed harmful to us has had a similar selective effect on our psyche, favoring fear of the small over trust and curiosity?

Most of animal life on Earth is small. Over 90 percent of known species are smaller than a human finger, smaller, in fact, than your fingernail. Our perspective on reality is severely handicapped by our gargantuan size, rare giants surrounded by the smaller majority. Our enormous size prevents us from appreciating, or even noticing, most of what shares this planet with us and forces us to focus our attention on other equally large, or larger, creatures. We proclaim kinship with wolves and deer, even while we hold our breath before squeezing the trigger, and cultures across the globe revere eagles, bears, and lions, but few pay any attention to lizards and snails. Size is the great divider, rooted in our atavistic need to conquer and subjugate, and small things are more easily dismissed as not worthy of serious attention. Tiny creatures are lumped into general categories of bugs, creepy-crawlies, or vermin, and at best are described as "strange," "bizarre," or "alien." But we forget that humans are the most recent arrival on this planet, peculiar products of evolution who have somehow escaped the rules that govern normal species' distribution and population growth.

With large animals everything is simpler—direct observations and experience have taught us that tigers can kill and turkeys make a good meal. With small things we lack either the patience or the ability to make observations, and end up drawing false, often ridiculous conclusions. And because we do not understand small creatures, we fear them. People in Madagascar are afraid of chameleons even though they have lived surrounded by these innocent animals for thousands of years. In Zimbabwe I met highly educated men who dread harmless blue-tailed skinks, considering them extremely venomous.

North American and European cultures are no better, fostering beliefs that daddy longlegs are venomous or that milk snakes steal milk from cows.

Unlike most mammals, who live in a sensual world dominated by scents, ours is a species that relies on vision. Eyes help us make emotional connection with other people as well as other species. We prefer animals that can return our gaze, which puts many smaller organisms, some of which may have "too many" eyes or none at all, at a great disadvantage in the struggle for our affection. But even as a young child I suspected that the smaller the animal, the more fascinating it must be, even if you cannot look directly into its eyes. And I am sure it had nothing to do with the fact that I was the smallest boy in my class, with thick glasses and a nickname: "The Bug."

In 1988 I took a break from college in my native Poland, and after twelve grueling months of bartending, washing dishes, picking berries, and painting walls, I finally saved enough money to embark on my first tropical voyage, one that would take me first to countries of Southeast Asia and later to Africa. It was the beginning of a much longer journey that has covered six continents, one that I hope to continue until I am old and incapacitated. It is a journey both on the surface of our beautiful planet and in spatial dimensions tens and hundreds of times smaller than the world we live in. This book is a collection of my travel snapshots.

On its pages I attempt to celebrate everything that is small and misunderstood. I pointedly ignore organisms typically portrayed in popular natural history writing and photography. Thus, there are no birds and mammals here, and most organisms I have photographed would fit inside a matchbox. There are a few exceptions for larger organisms, such as caecilians or land crabs, included here because they too share the stigma of being "noncharismatic." An outpouring of my lifelong fascination with the smaller majority, this book celebrates the enormous diversity of life found right under our feet. It does not pretend to be an exhaustive overview of tropical biota, and its taxonomic coverage is fragmentary. Each page provides only a glimpse into an animal's world rather than a comprehensive account of its life cycle. I have tried to give organisms largely ignored by nature writers their day in the sun. Thus I photographed, among others, ricinuleids, onychophorans, and fulgorids. I strive to show that katydids are more than just food for birds, and that flatworms can be beautiful. All photographs in this book were taken in the animals' natural habitat, often in remote, rarely visited locations. Some show organisms never before captured by a camera, and in some cases the animals are completely new to science.

I often fear that some popular media damage our perception of nature, particularly regarding our understanding of the smaller and, in the long run, far more important organisms. Watching natural history programs may leave some viewers with the impression that our unstable, volcanic planet is inhabited mostly by sharks, lions, and deadly snakes. I am afraid that this approach, which emphasizes the extremes of certain biological phenomena, dulls our senses and robs us of the ability to appreciate subtler and more beautiful aspects of nature. A child whose encounters with the natural world are limited to or mostly "documentaries" relating croc-attack survival stories is less likely to observe and be fascinated with the behavior of a chipmunk or understand why we should care

about flies pollinating wildflowers. Even some publications and museum exhibits that profess to promote our understanding of invertebrates and other small animals emphasize their "weirdness" and the danger they present, further widening the rift between the "higher" (large, warm, and cuddly) and "lower" (small and cold-blooded) animals.

In this book I try to restore some of the dignity of our smaller relatives by referring to them by their proper and scientifically accurate names. In preparation for this publication I contacted many specialists to ensure that all species portrayed here are correctly identified and descriptions of their biology and behavior are factually correct.

The book focuses on three major tropical ecosystems: humid forests, grasslands, and deserts. As any ecologist will tell you, this division is arbitrary. It serves merely to organize the pictorial essays in this book. All three ecosystems are terrestrial, although I included a number of organisms that bridge the gap between water and land.

In general, I omit any mention of an organism's size, so as not to detract from our appreciation of the animal's beauty or behavior. The images and stories here may not reverse the opinions of those with a phobia for invertebrates and other small animals. All I ask is that readers try to notice and understand them. Understanding is a prerequisite to caring, and caring is the key to saving.

Invertebrates and other small animals are disappearing far faster than their larger cousins. It is not easy to put precise numbers on this decline because it is significantly more difficult to document the extinction of a beetle than that of a bird. We lack baseline data on their distribution, and in many cases severely endangered species may not even be known to science. Yet, based on our knowledge of the population sizes and dispersal abilities of invertebrates and small vertebrates we can be certain that the loss of even a small patch of an old-growth tropical forest spells the demise to tens, perhaps hundreds of species. It will never be known how many species we have lost already with the forests of Madagascar or Java, but we can be certain that they were unique, beautiful, and irreplaceable. Still, documented extinctions have been recorded among snails, stoneflies, crayfish, grasshoppers, frog, toads, and many other small organisms. Results of the recently published Global Amphibian Assessment revealed a terrifying picture of nearly one-third of all species of frogs and salamanders being threatened with extinction, thirty-four species confirmed extinct, and an additional 134 species possibly extinct. A similar assessment is underway to evaluate the status of dragonfly and fish species worldwide, and I am afraid that its results may be even more frightening.

Current conservation efforts are by and large skewed toward large, warm-blooded, "charismatic" organisms, often dismissing insects and other invertebrates as "bird food." There is a certain rationale behind this approach—large, appealing "umbrella species" such as gorillas or rhinos help protect habitats for many other, smaller creatures. But the science behind this approach is not always sound. The Siberian tiger is certainly charismatic, but it is not a good umbrella species—by protecting it and its habitat we save probably less than 1 percent of species we would have saved if we used as an umbrella species the katydid *Lonchitophyllum,*

which can only survive in undisturbed, natural forests of Madagascar. An alternative approach to preservation of individual species or populations is the protection of hotspots of biological diversity, areas that combine the highest concentration of unique species and are under immediate threat of destruction. These hotspots are identified by combining all available data on plant and animal distribution, including the smaller and seemingly unimportant groups, such as termites or frogs.

From pollination to seed dispersal, from soil production to waste removal, and from water filtering to being food for others, invertebrates make Earth a livable planet. As tragic and unforgivable as it would be, the disappearance of mountain gorillas would have far smaller ecological repercussions than the extinction of a single species of savanna termite. We should never have to choose between these two species, of course. It is a very encouraging sign that both conservation practitioners and the general public are beginning to realize how important the smaller majority is to the health of our planet. In recent years, more resources have been directed toward documenting and protecting smaller organisms. Public appreciation of the beauty and importance of these animals is our strongest ally in this conservation work.

I hope that the images in this book will reinforce a child's interests in the natural life of caterpillars or frogs, or perhaps they will awaken a long-forgotten fascination with small creatures in an older reader. It may even encourage some to kneel down, look closer, and discover the beautiful world around our feet.

A Costa Rican earwig
(*Spongiphora* sp.)

TROPICAL

HUMID

FORESTS

S TEPPING OFF THE PLANE in Bangkok, I took a deep breath. The air was heavy, humid, and warm. Although I could smell airplane fuel, there was something else in the air that overpowered any manmade scent. It was a mixture of crushed leaves, some mysterious flowers, a sharp yet pleasant whiff of mold, and a hint of wet earth after the rain. It was the beginning of an addiction.

Between the confines of the imaginary lines of the Tropics of Cancer and Capricorn lies an area where life seems to run in a higher gear. Most of this area is or at least until recently was covered with forests. With a few exceptions, forests of the tropics are rainforests. Although precise definitions of tropical rainforests vary, most characterize these unique plant communities by a combination of intense precipitation regimes and high, relatively stable temperatures. Tropical rainforests are usually restricted to elevations below 1,000 meters, where the mean temperature of the coldest month does not fall below 18°C, and the rainfall exceeds 2,000 millimeters per year. Such conditions, combined with a short or completely absent dry season, allow for the development of a closed canopy, where at least 85 percent of trees are evergreen. The canopy of a true rainforest is situated at a height of at least twenty-five meters, with individual trees, so-called emergents, protruding above it as high as fifty meters. Within these forests, plant and animal diversity reaches levels unparalleled by any other ecosystem on the planet. Precise numbers are impossible to conjure, but for many groups of plants and animals over 90 percent of their species occurs nowhere else. Plant diversity in the rainforest can be staggering, with individual sites having more species of trees than the whole of Europe or the United States. Bird and mammal faunas show similarly spectacular patterns, but it is within the realm of invertebrate animals that tropical rainforests show the true meaning of the word diversity. Not only can a single site have more species of katydids or beetles than entire northern continents, but a single *tree* at a lowland tropical location can have more species of ants than many countries in Europe or regions in the United States.

Where do all these animals come from? Why are rainforests so incredibly diverse—saturated, it seems, to the very limits of their capacity with unique life forms? There are almost as many theories, some contradictory, as there are tropical ecologists, but they can be divided into two groups, one favoring explanations based on biotic and the other on abiotic factors. Higher structural complexity provided by plants of tropical forests may account for a greater number of potential ecological niches to be exploited by animals, resulting in more species having their own, albeit often very narrow, living space. Along similar lines, some claim that interactions are more intense in the tropics. They base this notion on the assumption that the favorable climate of the rainforest results in species living close to their carrying capacities, i.e., maximum sustainable population sizes. This forces them to compete for more and more finely divided resources, which results in a narrow specialization of each species. An exactly opposite view is voiced by the proponents of the predation theory, which assumes that species live at lower than theoretically possible carrying capacities, curtailed by the abundance of predators and parasites. Consequently, this lower than possible abundance of individual species makes more resources available to others.

Abiotic theories explain the exceptional biodiversity of tropical forests with inanimate causes. Greater amounts of solar radiation reaching the tropical belt of Earth results in greater primary production of plants, which translates into more species being able to feed on them and thus coexist. Another outcome of the high amounts of energy in the tropics is that organisms can mature and reproduce faster, producing more generations than organisms in colder climates. More generations means more genetic mutations natural selection can act upon, leading to more distinct genomes and species. Historical reasons may also be responsible for the high biodiversity of the tropics. Since the tropical belt of Earth has never been directly subjected to the devastating effects of glaciation, its physical conditions have remained unchanged for much longer than comparable areas closer to the polar regions. This gave the organisms living within it more time to evolve. Lastly, and most simply, because Earth is a sphere, the tropical belt is the largest climatically uniform area of the planet, allowing for more species to live there.

It is certain that there is more than one reason for exceptionally high tropical diversity, but regardless of the cause, the end result is an ecosystem teaming with finely spaced, intricately interrelated, and co-dependent organisms. The complexity of interactions among rainforest inhabitants often appears to have no limits, and it may be safe to say that if you think that you know the true nature of a particular relationship between species, then you have not studied these species carefully enough. Even seemingly straightforward affiliations such as between a plant and its pollinator may involve additional players that can help or harm this relationship. A good example of the complexity of relationships among rainforest organisms and the opportunistic nature of many of them is the microecosystem of a Peruvian bamboo *(Guadua weberbaueri)*. The stem of this plant is divided into short, contained nodes, which over time fill up with water secreted by the plant. These tiny, ready-made pools are inaccessible to animals until a female katydid *(Leiobliastes laevis)* makes an incision in the wall of the bamboo stem while laying her eggs there. The incision eventually erodes, allowing other organisms to enter the node. In the end, the node becomes occupied by a community consisting of nearly thirty species of animals, including flies, katydids, damselflies, beetles, and earthworms, many of which can breed and develop only there.

Some say that a biologist in the rainforest is like a kid in a candy store, and I certainly feel this way every time I am there. Standing motionless on a path crossing the forest, I can sense the river of life flowing all around me, each square inch of my surroundings vibrant with activity. Life in the rainforest seems unstoppable, and even death loses some of its finality when witnessed in the context of perpetual rebirth, regrowth, and recycling. Trees fall, creating large gaps that spur growth, allowing new species of trees to reclaim their places in the forest. The forest heals itself, changing over time its species composition, and the balance of dominance of certain organisms invariably tips in favor of others. It is clear now that because of the extremely heterogeneous nature of the rainforest, the disappearance of a single species has little effect on the ecosystem as a whole.

But there are limits to how much change a forest can take. Human encroachment is something that these ancient ecosystems cannot withstand, and the devastating efficiency with which we ransack tropical forests leaves little hope for their long-term survival. It took us a very long time to realize that a forest is not just a sum of its trees. Forests, especially those in the tropics, are as much responsible for creating hospitable conditions for human survival as are the ozone layer and fresh water. But the impact of problems in our daily lives often outweighs and displaces our concern for a planetary disaster brewing in another hemisphere. A five-dollar increase in our monthly cable TV bill probably makes us angrier than the loss of a thousand hectares of rainforest in Indonesia. The effects of habitat loss must strike us directly to make any difference in our perception of environmental problems. It took massive mud slides and hundreds of deaths in the Philippines for the politicians in this country to finally grasp the connec-

tion between uncontrolled deforestation and the steadily more devastating effects of tropical storms. In an unprecedented move, all logging was recently suspended in the Philippines. Hopefully, other nations will learn from the Filipino experience.

If you have not yet seen a tropical rainforest, do it now. Plan your next vacation as close to the equator as possible. Your visit to a nation that still has this spectacular ecosystem will prove its value as more than just a source of timber and may ultimately be the key to its survival. Nations such as Costa Rica realized the value of the rainforest as a fantastic attraction to nature lovers worldwide, and are now reaping economic benefits from their ever-growing ecotourism industry. There are countries like Gabon or Peru that can still follow a path of economically profitable conservation, and it is largely up to citizens of other countries to make this possibility a reality. I know I will be doing my part. The sweet scent of the tropics has got me hooked.

Quickly vanishing forests of West Africa are still home to one of the most magnificent members of the beetle order—the Goliath beetle *(Goliathus regius).* Despite their bulky appearance, Goliath beetles are excellent fliers, frequenting flowers blooming in the forest canopy. Adult beetles feed also on ripe fruits and sap, while their giant grubs, which can exceed 150 millimeters in length, develop in the decomposing wood of naturally fallen large trees. It is the increasing shortage of large rainforest trees that may ultimately spell the demise of this and other giant beetles. Their development depends on the availability of old-growth trees, the very same ones that are the principal target of the logging industry.

J. B. S. Haldane, the British geneticist, is often quoted for proclaiming, "The Creator, if He exists, has a special preference for beetles." With between 275,000 to 350,000 described species of beetles it is hard to argue with this statement. Of these, about 60,000 species belong to a single family, the weevils (Curculionidae), making it not only the largest among beetles, but also of all living organisms. Their evolutionary success may be based in the morphology of their mouthparts, which in most species are situated at the tip of a long rostrum or snout. This allows such species to bore deep holes in even the hardest seeds or nuts in order to deposit their eggs there, providing developing larvae with a safe environment and a rich supply of food. It is also certain that the weevils' success is related to the diversification of flowering plants, which began in the Cretaceous. A process known as coevolution, which can be compared to an arms race between plant-feeding beetles and their hosts, has led to a steadily increasing specialization of beetle species in response to the evolution of better defensive strategies of plants, and an almost exponential increase of their numbers.

The bearded weevil *(Rhinostomus barbirostris)* is a species with an interesting sexual polymorphism. A proportion of males in each population is smaller than other individuals of this sex, and resembles females in their appearance. This allows them to sneak unnoticed past the larger males and approach the female without being challenged (Costa Rica).

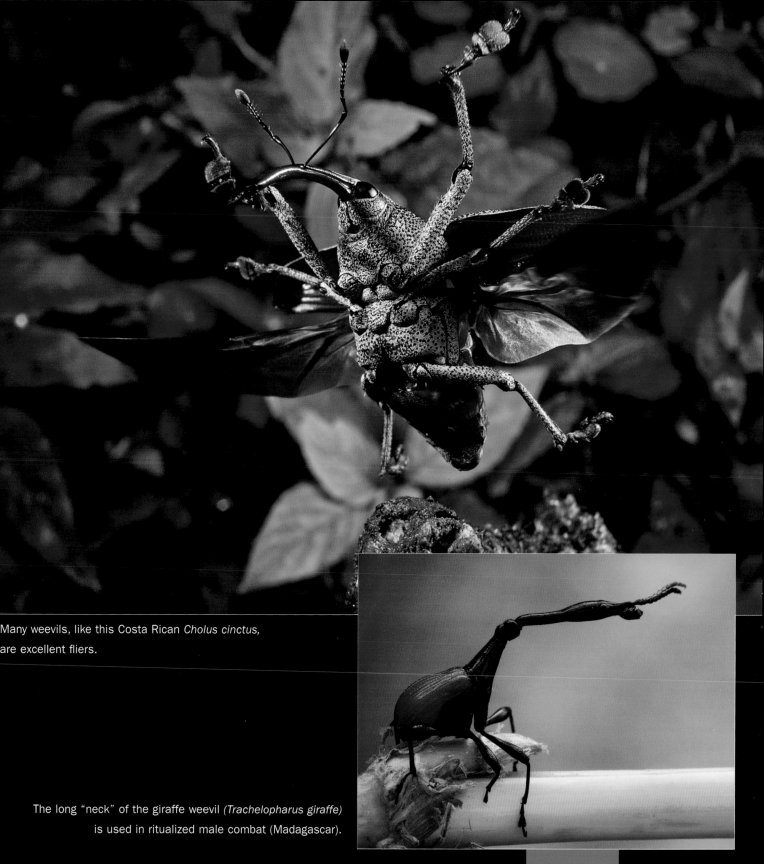

Many weevils, like this Costa Rican *Cholus cinctus*,
are excellent fliers.

The long "neck" of the giraffe weevil *(Trachelopharus giraffe)*
is used in ritualized male combat (Madagascar).

15

Longhorn beetles (Cerambycidae) include some of the most beautiful insects to be found in a tropical rainforest. True to their common name, their antennae are exceptionally long and thick and are significantly longer in males than in females, being used to detect sex pheromones produced by females or aggregation pheromones produced by conspecific males. Both sexes of at least some species are also attracted to volatile chemicals emitted by their host plants. All members of this large family develop within plant tissues, boring deep corridors in the wood of trees or stems of herbaceous plants.

Many adult longhorn beetles stridulate loudly if disturbed or captured by a predator, using a washboard-like surface on the back of their head that they rub rapidly against the edge of the "neck" region, or the pronotum. The sound produced by these beetles has a wide frequency spectrum that resembles the sound produced by potentially dangerous or noxious organisms, such as rattlesnakes or cockroaches equipped with chemical defenses. It is unclear if the beetles' behavior is an example of acoustic Batesian mimicry, in which harmless organisms mimic potentially dangerous ones, or a genuine warning. Longhorn beetles have exceptionally strong mandibles, which they can use to can inflict serious wounds to a careless predator.

A slender *Taeniotes scalaris* prepares to take off (Costa Rica).

An unidentified species of longhorn beetle (Madagascar)

The longhorn beetle *Sternotomis pulchra imatongensis* from Guinea is a diurnal species with bright coloration.

An unidentified longhorn beetle (Guinea)

The metallic coloration often seen in rainforest insects is the product of light diffraction in the microstructure of their cuticle. The function of such splendid colors is not entirely clear in all organisms that display it, but the most likely function is that of crypsis. Whereas a metallic green insect looks exceptionally conspicuous if placed on a matte surface, it blends seamlessly among wet, highly reflective vegetation in the humid forest. The smooth surface of some insects' cuticle acts as a mirror, changing its hue depending on the surroundings. Jewel beetles (Buprestidae) sometimes mix the highly reflective surfaces with dull patterns created with a waxy secretion of their cuticle, thus breaking up the shape of their bodies into less recognizable elements.

Jewel beetle (Euchroma gigantea) (Costa Rica)

19

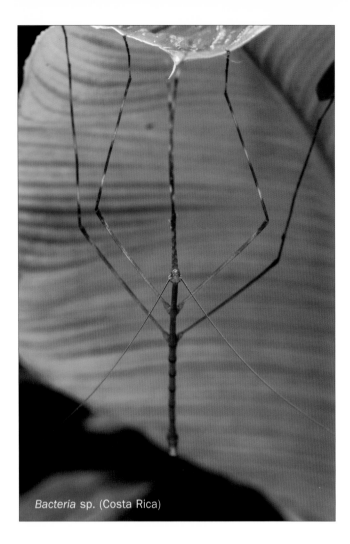

Bacteria sp. (Costa Rica)

Metriophasma diocles (Costa Rica)

Eggs of Eurycnema goliath (Australia)

The term usually associated with walking sticks (Phasmatodea) is "mimicry." There is no doubt that these insects have perfected the art of pretending to be plants, even their eggs are such superb replicas of plant seeds that they can fool a botanist. But walking sticks have a second and third line of defense. Startle display is common among species with fully developed hind wings (the first pair of wings is nearly always reduced in these insects), and frequently these are strikingly colored. Sometimes, as in Metriophasma diocles, the wing has a darker or differently colored spot, perhaps suggesting an eye of a larger animal. But in addition to visual defenses, many walking sticks employ an arsenal of chemical weapons. Little is known about the chemistry of most tropical species, although many such as members of the Neotropical genus Prisopus produce a strong odor if disturbed, and are clearly avoided by predators. A unique toxic compound, anisomorphal, was discovered in some highly noxious walking sticks. Unlike grasshoppers and other insects who feed on toxic plants and sequester their secondary compounds, walking sticks can feed on a variety of innocuous plant species and synthesize their own chemical protection.

Prisopus sp.
(Costa Rica)

Coconut crab *(Birgus latro)*

A Conquest of Land:
Terrestrial Crustaceans

THERE IS LITTLE DOUBT that life on our planet originated in an aquatic habitat, most likely in the primordial ocean covering most of its surface about 3.8 billion years ago. The evolutionary history that followed witnessed repeated escapes from water, with different degrees of water independence achieved by now-terrestrial organisms. Some groups such as early arthropods achieved a remarkable success in the form of explosive radiations among their land-dwelling descendants and complete independence of aquatic life. Others, like flatworms (Platyhelminthes) or roundworms (Nematoda), were never able to develop efficient water retention mechanisms and can only survive in extremely humid environments, such as wet soil or as parasites inside the bodies of other organisms. Between these two extremes are animals that can survive almost their entire lives on dry land, yet in order to reproduce and develop must return to water. One such group is the crustaceans.

Different lineages of aquatic crustaceans have independently attempted to colonize land several times. By far the most successful were the sowbugs and pillbugs. About 5,000 species of these multilegged animals, which are about the size of a fingernail, are members of the crustacean order Isopoda. They are able to live and reproduce on land, often in remarkably dry environments, but in order to achieve this they had to overcome two major obstacles. Being the direct descendants of strictly aquatic animals, their respiratory system was adapted to breathing under water, requiring constant submersion in order to extract oxygen. Evolution overcame this challenge by hiding the gills of terrestrial isopods inside some of their modified legs. In effect, their breathing system became remarkably similar to, yet independently evolved from, that of insects and arachnids. A simplification of their reproductive cycle, eliminating the free-living aquatic larva, allowed them to radiate into drier habitats. By completing their early development inside an egg protected by the female, pillbugs and sowbugs were able to cut all remaining ties to the ocean. Thanks to these two major evolutionary inventions, in some habitats they can successfully compete against the unquestionable rulers of the land—insects. But while Isopoda are the only truly terrestrial crustaceans, visitors to the tropics rarely see them. Instead, a walk on the beach or even a casual stroll in the forest may lead to an encounter with a much bigger terrestrial convert: a land crab.

Crabs, these quintessentially marine creatures, ventured onto land many times during a long evolutionary history that dates back to the early Jurassic period, over 180 million years ago. Some were able to colonize land by first adapting to freshwater habitats, on its own a considerable challenge to any organism coming from the ocean. The more ephemeral, by geological standards, nature of lakes and rivers eventually drove some species to dry land, and many are capable of surviving far from permanent bodies of water. Instead, their larval development relies on tiny, temporary pools within epiphytic plants in the rainforest or water accumulated in tree holes, or their development takes place in an egg attached to the mother's body.

Other crabs were able to leave the ocean by a more direct route. Fiddler crabs (*Uca*) and ghost crabs (*Ocypode*) represent an intermediate stage in the process of land colonization, and it would be difficult to classify them as either truly terrestrial or fully aquatic animals. Although fiddler and ghost crabs spend their adult lives on land at the very edge of the ocean courting mates and looking for food, they must return to water frequently to moisten their gills. Oceans are full of specialized predators that target crustaceans, and by foraging on land these crabs are able to escape them. Of course, seagulls and other predators on land will gladly devour an unsuspecting crab, but crabs' excellent vision and extreme agility make fiddler and ghost crabs difficult targets.

Some marsh crabs (*Sesarma* and *Armases*) were able to move much farther away from the sea. Although they are still limited by their aquatic larval development, thanks to some extraordinary physiological adaptations they are able to complete it in very small amounts of water. For instance, some species develop in overheated and oxygen-depleted tidal pools on rocks at the edge of the ocean, while others complete their larval stages in water accumulated on epiphytic plants high in the trees. In one species of *Sesarma* the female finds an empty shell of a terrestrial snail, fills it with water, and lays her eggs there. These innovative reproductive strategies allowed marsh crabs to colonize many land habitats, including the rainforest canopy. Most species are scavengers or predators, hunting smaller invertebrates, but a few switched to a vegetarian diet. One Brazilian species of *Armases* feeds on flowers of hummingbird-pollinated plants and effectively competes against the birds trying to access the nectar.

But the largest and most widely distributed land crabs belong to the family Gecarcinidae, or true land crabs. Like most other terrestrial crabs, they employ the so-called "export strategy" in their reproduction—adults spend their whole lives on land, so larval development takes place in the dramatically different marine environment. In fact, the adults can no longer survive in the ocean, and will drown if submerged for too long. Because adult crabs may live many miles inland, their breeding is synchronized, often using phases of the moon as their cue to start a migration toward the shore. This assures that all reproductive individuals will be present at the edge of the water at exactly the same time, making finding a partner fast and easy.

True land crabs are well known to people living along tropical coasts across the globe. Many species are very large, weighing up to one and a half pounds, and in many countries their meat is considered quite a delicacy. The great blue crab (*Cardisoma guanhumi*), once common in Florida and most Caribbean islands, has been hunted to near extinction in many places, and is now a protected species in parts of its range. Species of *Gecarcinus* are similarly sought for their meat. I always liked its taste, but after seeing these animals twitching in the sun at a fish market, still alive but with all their legs torn off and sold separately, I decided to eschew crabmeat.

Overharvesting is a serious threat to many populations of land crabs, just as it is to their marine cousins and other crustaceans. There is a tendency within almost every commercially exploited species of these animals for the average size of an individual in a population to shrink. It happens not because of a genetic trend, but because the largest ones are always the first taken. Fortunately, in many instances properly introduced and enforced regulations have helped local populations rebound to their former sizes.

Another serious danger that threatens the survival of some species of land crustaceans, one that is far more difficult to control, is the introduction of non-native species of animals. Rats and pigs rampage on many islands where land crabs used to have few natural enemies, destroying their habitats and decimating their population. One particularly dramatic example of an exotic species becoming a destructive force for a local crab population comes from Christmas Island. This small volcanic speck of land off the coast of Java is home to one of the most spectacular examples of animal migration. Every year in November millions of large, bright red crabs (*Gecarcoidea natalis*) start a march from the forests in the interior of the island to the sea where they spawn. There used to be so many crabs on the island that they became a natural force, shaping the tree composition in the forest and moving tons of soil. At one point they were also able to protect the island from the invasion of an introduced and very destructive African snail (*Achatina fulica*), crushing its thick shell with their powerful claws. But their mighty muscles and hard exoskeleton were no match for the tiny but vicious ant *Anoplolepis gracilipes*, accidentally introduced at the beginning of the twentieth century. These ants are becoming increasingly more common in the rainforest favored by the crabs, where they attack and kill them only to make a nest out of their victim's burrow. Controlling the ants is now the priority in the conservation efforts to save the crabs. More recently, the same species of ants decimated the population of ghost crabs in the Seychelles. Ports of entry across the globe now have significantly more efficient screening processes, and stricter regulations of the movement of livestock and plants will hopefully reduce the risk of similar biological disasters happening elsewhere.

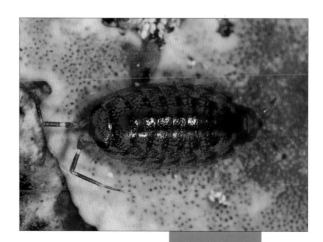

(Clockwise from upper left) Fully alert, a male ghost crab *(Ocypode gaudichaudi)* scans the horizon for mates and rivals (Costa Rica, Pacific coast).

First to discover a dead fish, a young ghost crab *(Ocypode quadrata)* hits the jackpot (Costa Rica, Atlantic coast).

Close relatives of ghost crabs, fiddler crabs *(Uca)* feed on organic matter suspended in mud. Males have an enormous claw, which they use to attract mates and ward off competitors (Dominican Republic).

Nothing can escape the excellent vision of a ghost crab *(Ocypode gaudichaudi)* as it patrols a beach on the Pacific coast of Costa Rica in search of small animals, carcasses of fish washed ashore, or any other edible organic matter. It will not shy from attacking other crabs, and for this reason younger individuals make their burrows away from areas frequented by older and larger members of their species. Though largely terrestrial, ghost crabs have gills that need to be moistened frequently. Their deep burrows reach below sea level, allowing them to submerge their gills without risk of being swept away by the waves.

Ghost crabs employ a complex language of visual and acoustic signals in their daily lives, and some species use three different types of sound to attract their mates and mark their territories. They have earned their name thanks to an uncanny ability to disappear in the sand after a blazingly fast dash away from a predator or beachcomber. But like many other terrestrial crustaceans, in many places they fall victim to introduced ants *Anoplolepis gracilipes,* which attack the crabs in their burrows and kill off entire populations. These crabs are also very sensitive to the damage done to beaches by human recreational activities and have disappeared from many places where they used to be common. Thanks to this sensitivity, some species of ghost crabs have been used as reliable indicators of the impact people have on sandy beaches.

Unable to break her ties to the sea, a female blue land crab (*Cardisoma guanhumi*) cautiously approaches the edge of the beach to release her eggs during the full moon. She cannot swim, thus she must be careful not to be swept away by the waves, and soon she runs back to her burrow in the forest. Her planktonic larvae will develop into tiny crabs in less than two months and then will leave the ocean to begin terrestrial life.

Bright coloration combined with a pair of powerful claws make for a convincing threat display in this red land crab *(Gecarcinus ruricola)* (Costa Rica).

Males of the blue land crab sport giant claws used in territorial combat (Dominican Republic).

Most hermit crabs spend their lives at the bottom of the sea, but members of the family Coenobitidae left the water in favor of the terrestrial environment. Untrue to their common name, most terrestrial hermit crabs are highly gregarious, and thousands of them can be seen on tropical beaches scavenging for pieces of coconut and sifting through flotsam and jetsam. Some species, like this *Coenobita clypeatus,* can be found many miles from the ocean, even in such unlikely settings as mountaintops on some Caribbean islands. The abdomen of the hermit crab is soft and vulnerable and must be protected in an empty snail shell. The availability of suitable shells is a major factor limiting the distribution of hermit crabs, and many species do not venture far from the beach for risk of not being able to find one. Yet on many Caribbean islands large and thick shells of the land snail *Caracolus* provide an excellent replacement for shells of sea snails carried by younger individuals, and large, adult hermit crabs are common in the litter of inland forests.

A young hermit crab *Coenobita clypeatus* cautiously emerges from its shell after being hit by an incoming wave on a rocky beach of Hispaniola. Its giant claw fits perfectly into the opening of the shell and effectively protects the more vulnerable parts of the crab's body.

One of my lifelong dreams had been to see a live coconut crab. It was finally fulfilled on the Pacific island of Guadalcanal. The coconut crab *(Birgus latro)* is not just another land crab—it is the largest living terrestrial invertebrate, reaching a weight of nine pounds and a leg span of over three feet. Their lifespan is equally impressive, and the largest individuals are believed to be forty to sixty years old. Contrary to their name, coconut crabs eat more than just coconuts and their alternate name, robber crabs, reflects their feeding behavior far more accurately. They are not afraid to use their formidable size and strength to chase away other crabs or rats feeding on carrion or other desirable delicacies. They have been observed hunting frogs, young turtles, and even terns and other seabirds. Recent observations confirm the long-held belief that coconut crabs are capable of stripping the husk off and opening coconuts. They have also been seen cracking the extremely hard shells of macadamia nuts. They are superb tree climbers and will try to hide on palm trees if cornered.

Coconut crabs begin their lives in a way very similar to their closest relatives, hermit crabs. A planktonic larva spends three to four weeks in the ocean before emerging on the shore of one of the tropical islands of the Pacific or Indian Oceans as a miniature hermit crab carrying a snail shell. It soon abandons the shell and begins a solitary life. Coconut crabs can be very aggressive. Upon encountering another individual of its own species, it engages in a ritualized display that involves lifting its legs to demonstrate its superior size. Coconut crabs do not hesitate to attack smaller individuals, and cannibalism is a major selective force regulating population sizes of this species.

One of the drawbacks of living on land is the abundance of insect parasites. Parasitic flies can attack terrestrial crustaceans, as shown by these fly pupae attached to the underside of the abdomen of a coconut crab.

An arboreal
crab (*Malagasya
antongilensis*) guards
a tree hole in a remote
montane forest of
northern Madagascar.

A female of
*Liberonautes
latidactylus* scouts
the forest floor near a
small stream in Guinea.

Several groups of crabs managed to sever their ties to the sea and complete their development on land or in small reservoirs of water collected in tree holes or epiphytic bromeliads. Marsh crabs, like this female of *Armases* sp., belong to the family of marine rock crabs (Sesarmidae), but thanks to innovative physiological adaptations, they can survive and breed far from the sea, including the rainforest canopy. This species is common in the trees of coastal forests of Costa Rica, where it hunts beetles and other insects.

Finding a live velvet worm is every zoologist's dream. Considered to be one of the relict groups of animals, modern velvet worms (Onychophora) appear to be remarkably similar to fossil forms dating back as far as the Cambrian 530 million years ago.

Their current distribution, restricted to areas once part of the ancient land mass Gondwanaland, supports their prehistoric provenance. They are now found only in Australasia, southern Africa, and parts of Central and South America. There they can be found in humid, warm habitats, such as decaying logs in the rainforest or on moss-covered tree trunks.

Almost every aspect of the velvet worms' biology makes them stand out among other animals, beginning with their reproduction. Mating in velvet worms involves a gamut of sperm delivery mechanisms. The simplest one involves deposition of a spermatophore directly on the female's skin, which dissolves underneath and allows the sperm to enter her body. The sperm eventually find their way to the ovaries. In some species the male deposits sperm on top of the female's head, whereas in others it is the male who puts his head into the female's genital opening at the rear end of her body to deliver sperm. Their breeding is equally diversified, with some species laying eggs, others allowing the eggs to mature and hatch inside the female's body, and still others forming a placenta and giving birth to live young, bypassing the egg stage entirely.

Velvet worms are predominantly predaceous, feeding on small insects, spiders, and crustaceans. Being slow and soft-bodied, they rely on an unusual hunting technique, ejecting sticky, silk-like strands from a pair of oral glands. These strands harden quickly, but remain sticky thanks to the presence of the slow-drying surfactant nonylphenol, imprisoning the victim. The velvet worm then approaches and injects some of its digestive saliva. While the victim dies, the velvet worm eats the sticky strands used to ensnare the victim, recycling precious proteins used to produce them. Eventually, the content of the victim's body is sucked out using sickle-shaped mandibles.

The interest velvet worms elicit from zoologists is also their curse, and in Australia some species have been overharvested to fulfill the demand for specimens used in zoology classes at various universities. In other parts of the world, habitat loss is the main threat faced by velvet worms, although in South Africa some species have been able to adapt to living in disturbed habitats covered with nonnative vegetation. A Brazilian species, *Peripatus acacioi,* has the honor of being the first and so far only terrestrial invertebrate in Brazil to be used as a flagship species to inspire the creation of a nature reserve.

Frequent downpours, typical of most rainforests, blur the line between water and land, allowing some generally aquatic organisms to venture into new, drier habitats. Planarians (Tricladida) are the only members of the flatworms (Platyhelminthes), a group most diverse in oceans and lakes that managed to adapt to living in terrestrial environments. They move on land by the action of the "creeping sole," a strip of powerful, closely spaced cilia on the ventral part of their body. In addition, they exude mucus, which, in addition to smoothing the path of their locomotion, can be used in a way similar to spider silk to create a strand that allows the planarian to lower itself safely from a branch to the ground.

Land planarians feed on small invertebrates, including termites. To capture its prey, the planarian waves the anterior part of its body near a column of termites, and the insects are captured and entangled in a sticky secretion of the planarian's body. Other planarians prefer to feed on earthworms, and several exotic species of these animals have caused a noticeable decline of native earthworms after being accidentally introduced to Australia, the United States, Great Britain, and other countries. Once fed, planarians can survive many weeks without eating, relying on nutrients stored in the digestive epithelium, and, if necessary, they can digest their own bodies, eventually shrinking in size but able to survive a prolonged period of starvation.

The Costa Rican dragonfly *Gynacantha tibiata* awaits the end of rain to start hunting again.

Few groups of rainforest organisms have been around for as long as damselflies and dragonflies. The earliest members of this insect order are known from fossils dating back about 320 million years ago, and their morphology has hardly changed since. With the exception of a few species that can develop on land in very humid habitats, all dragonflies and damselflies spend the first half of their lives in an aquatic environment. Both larvae and adults are predators relying on their excellent vision to capture prey. The aquatic larvae are equipped with long, protractile mouthparts, which can be shot out in a millisecond to grasp a snail, tadpole, or even a small fish. Dragonfly larvae are some of the few organisms that use jet propulsion in their locomotion—they can shoot a stream of water at a high speed from the end of their abdomen, swiftly forcing the animal forward.

Adult dragonflies and damselflies hunt small insects, including mosquitoes. Nothing brings me more pleasure than the sight of a dragonfly circling my head in the rainforest because I know that it will protect me from at least some bites.

Many dragonflies and damselflies are threatened by water pollution and the loss of their habitats. Nearly 140 species of these insects have been placed on the World Conservation Union (IUCN) Red List, and some of them are considered critically endangered.

Even a fierce predator such as this blue dragonfly, *Palpopleura portia,* can fall victim to an even larger hunter, in this case a preying mantid *(Polyspilota aeruginosa)* (Guinea).

The gorgeous damselfly *Sapho ciliata* is a common species in West African forests (Guinea).

Sierra Leone reed frog (*Hyperolius chlorosteus*)

Amphibians

I WAS A SMALL CHILD—at the age of twelve I still looked eight—and perhaps my minuscule stature was one of the reasons I felt a stronger affinity with the smaller citizens of the animal world. Frogs were some of the first animals with whom I truly fell in love. When I was about ten my fascination was so intense that nearly everything I remember from that period is in some way connected to frogs. Living in a big city made my encounters with frogs rare, and often I had to console myself with watching them in captivity. I remember racing past enclosures with big cats and antelopes in our city zoo, my mother calling me to come back right this minute, as I rushed to find the terrarium with the gorgeous green tree frogs. I also recall the joy of waking up in the morning to find a jar with a fire-bellied toad my father had caught for me the night before while visiting relatives in the country. As a child I did not accept the possibility that some did not share my fascination with amphibians, and the repulsion that some of my parents' friends struggled to hide when I paraded a jar with a toad or tadpoles in front of their faces was absolutely beyond comprehension. Naturally, I pledged that I would become a herpetologist, a biologist working on amphibians and reptiles. In the end I chose the path of an insect biologist, once those even smaller and more amazing animals captured my imagination. Yet I will always have a deep love for frogs and all their fascinating relatives.

Amphibians, a group that includes frogs, salamanders, and worm-like caecilians, are descendants of the first vertebrate animals to conquer land some 350 million years ago. They still carry the baggage of aquatic development, and most must return to water to breed and grow. A few have evolved ways of circumventing this requirement by either completing their development inside their mothers' bodies or laying eggs in moist habitats and emerging from them fully formed. Some go as far as completing their development in their parents' stomachs or in folds of skin on their parents' backs. Other amphibians went in the opposite direction and returned to water. Many salamanders spend their entire lives underwater, as do some caecilians and a few frog species.

Considering the close affinity of most amphibians with water, it is not surprising that the highest diver-

African clawed frog *(Silurana tropicalis)*

sity of these animals is found in warm, wet depths of tropical forests. For example, the tiny but hot and humid country of Costa Rica has 174 species of amphibians, compared with 193 species from the whole of North America north of Mexico.

In my visits to rainforests around the globe in search of katydids and other insects, I was fortunate enough to meet with herpetologists studying frogs and other amphibians in the area. Because most frogs, like most katydids, are chiefly nocturnal, we spent many nights searching for them in swamps and tangles of the forest. And because both animals communicate with sounds, we often made recordings of the same nocturnal choruses, one of us hearing only the bell-like tinkling of tree frogs, the other fascinated by a buzzing repertoire of a cone-head katydid. One person in particular, Dr. Mark-Oliver "Mo" Rödel of Würzburg University, has shown me some of the most fascinating albeit not the most colorful examples of frogs in the West African forest, including the fat and ungainly yet strangely beautiful *Silurana* and the live-bearing toad *Nimbaphrynoides*. In what follows I share some of the most beautiful or interesting examples of amphibian life I have had the opportunity to encounter.

These days frogs and other amphibians around the world are not doing well. Dramatic losses of their habitat, a fate shared with virtually all wild organisms, are not the sole cause. Frogs are falling victim to a global epidemic of fungal diseases. The culprit, a pathogenic fungus named *Batrachochytrium dendrobatidis* that belongs to a class of fungi known as chytrids (Chytridiomycetes), has spread like wildfire among frogs in Australia, Europe, and South, Central, and North America. Many species are al-

ready feared extinct because of the deadly infection; others may face this fate very soon. The fungus, which possibly originated in Africa and was subsequently introduced to Australia and other places, attacks the skin of frogs, or more specifically, a vital component known as keratin. In humans and other mammals, keratin is the main building block of hair and nails, but in amphibians it forms a protective layer in their otherwise soft, permeable skin. Once infected, the animals probably become more susceptible to the negative effects of ultraviolet light, pollution, or other parasites, and as a result, invariably die. The fungus also attacks keratinized mouthparts of tadpoles, making it impossible for them to feed. The tragic part is that this disease affects with equal force populations in areas of heavy human impact and those that are absolutely pristine, such as cloud forests of Central and South America. Scientists are working now on antifungal treatments that can be applied to populations of particularly heavily impacted species, but for many it may be too late.

The global decline of amphibians is further exacerbated by other factors, most notably pollution runoffs from agricultural and industrial areas and global climate change. The former produces the widely publicized deformities in frogs, causing them to grow extra limbs or eyes, and generally lowering viability of their natural populations. Stricter pollution regulations, if properly enforced, would help reverse this trend. But global warming, caused by high levels of carbon dioxide and other greenhouse gases from industries and deforestation, combined with depletion of the ozone layer, can easily wipe out entire species. Particularly impacted are frogs in high elevation areas, which are adapted to cooler, more humid conditions. Rising temperatures cause warm air to envelop their habitats, killing all organisms unable to cope with drier, warmer conditions. The recent Global Amphibian Assessment conducted by IUCN and Conservation International has found that as many as 1,856 species, or 32 percent of the entire world fauna of frogs and their relatives, are threatened with extinction. Can anything be done to change these very disturbing developments? As with most negative environmental impacts of human actions the answer is "yes," but only if we stop the accelerating pace of habitat degradation and implement recovery plans for areas and populations. Are we willing to do this? Current projections of human population growth predict a steady increase of our numbers, to peak at about 12 billion in 2050. This doubling of the human population will require dramatic increases in production of food and other commodities. And what is easier than converting areas unpopulated by humans, such as the vast forests of the Amazon, into beef pastures and sawmills? Even if current potentates of countries lucky enough to possess large, unspoiled areas agree to protect them, who is to say what will happen if the populations of these countries are faced with a choice of protecting the last stand of endangered species or feeding themselves? And yet habitat protection, rooted in precise, scientifically sound estimates of species composition

and their sustainability in a given area, is the best solution to species extinction we have now. In some cases protection of natural populations can be augmented with captive breeding programs, and in the case of some amphibians this may be the only way out, especially if wild populations are already gone or infected with deadly fungi. The balance between a comfortable human existence and preservation of Earth's natural heritage is increasingly more difficult to achieve, yet the optimist in me still has a strong voice. I believe that fifty or a hundred years from now a child's heart will jump at the sight of a tree frog discovered after the rain in her backyard, and a scientist in South America will solve the mystery of reproduction of yet another species of a caecilian.

Looking at a caecilian, it is hard to believe that this earthworm-like animal is actually a vertebrate, closely related to frogs and salamanders. Most species, like this *Gymnophis multiplicata* from Costa Rica, are almost exclusively subterranean, rarely emerging to the surface. In contrast to the majority of amphibians, caecilians do not need to return to water to breed. Their young hatch from eggs inside the mother's body and continue to grow there, feeding on nutritious secretions from the oviduct. Adult caecilians feed on earthworms and insect larvae.

Caecilians are blind, their eyes covered by bone or skin. A small bump, or the chemosensory tentacle, situated where the eye should normally be, serves as the primary scent organ since the nostrils are closed while the animal burrows underground.

The head of a caecilian has an extremely compact, solid skull, used by these animals to push through soil.

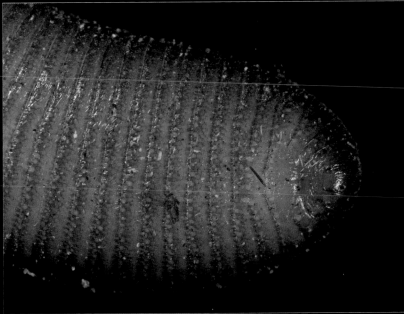

Unlike a snake or legless lizard, a caecilian does not have a tail, and the body ends abruptly in a cloaca.

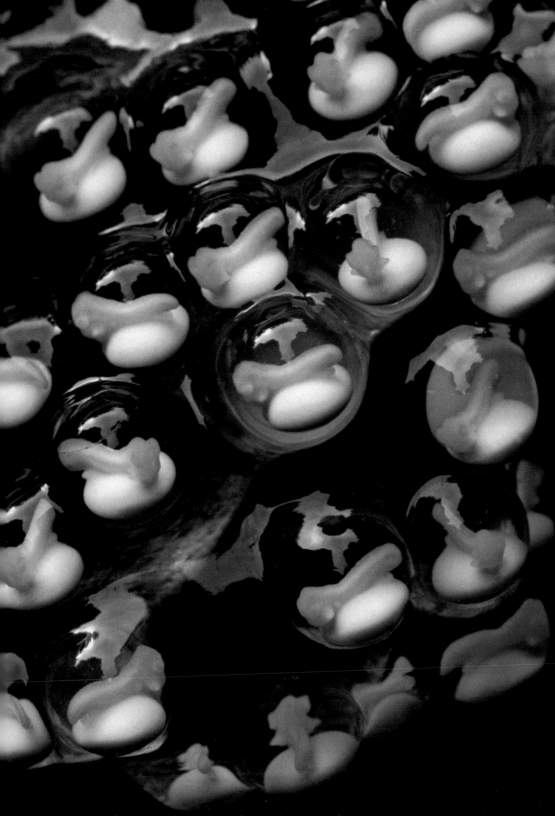

An ancient evolutionary heritage forces most frog species to return to water during breeding time. With few exceptions, tadpoles must spend their early lives as aquatic animals, feeding and breathing under water. But ponds and rivers are a perilous environment: fish and invertebrates devour both frog eggs and the developing larvae. Certain species of tree frogs (families Hylidae, Hyperolidae, and Rhacophoridae) have developed ways of reducing this risk by spending part of their initial development on land, where fewer predators target their eggs. In the Central American red-eyed tree frog *(Agalychnis callidryas),* both the courtship and egg laying take place on vegetation near or above small ponds. Eggs are attached to leaves where they begin their development, awaiting the arrival of heavy rains that will stimulate hatching of tadpoles and wash them into the water.

Since egg fertilization in frogs is external, the male must remain with the female until the very moment of oviposition to ensure his fatherhood. Often several males will try to fertilize the same clutch of eggs, and in some cases DNA fingerprinting has confirmed multiple paternity.

A few days old, each egg of the red-eyed tree frog is like a miniature aquarium, providing the developing embryo with shelter and, in the form a large yolk sac, enough nutrients to continue development and achieve independence (Costa Rica).

The male of the Malagasy web-footed frog *(Boophis luteus)* guards the female in a behavior known as amplexus; the actual fertilization of eggs will happen only when the female produces her eggs in a slow-flowing stream (Madagascar).

Ready to hatch, week-old eggs of the red-eyed tree frog with fully developed tadpoles inside begin to slide down the stem of the plant to which they were attached.

The large, swollen parotoid glands behind the head of the cane toad *(Bufo marinus)* produce a mixture of fourteen different toxins affecting the nervous system of any predator inexperienced enough to try to eat it. Thus, few animals prey on cane toads, and their distribution is larger than that of any other amphibian species in Latin America. They had also been introduced to Australia to eradicate a pest beetle species, but faced with no natural enemies and perfect breeding conditions, their population exploded there, causing considerably more harm than the species the toads were supposed to control.

Other toad species are not as fortunate, and many are seriously threatened. Whereas cane toads thrive in Costa Rica, their close cousins, golden toads *(Bufo periglenes)* from high-elevation cloud forests of Monteverde, have already gone extinct. A similar fate may soon meet another high-elevation species, the West Nimba toad *(Nimbaphrynoides occidentalis)* *(right)* from West Africa, an amphibian unique in its ability to give birth to live, fully formed young toads, an adaptation allowing it to survive in an environment with unpredictable availability of water.

Cameroon toad *(Bufo superciliaris)* (Guinea)
West Nimba toad *(Nimbaphrynoides occidentalis)* (Guinea)

51

Nearly invisible among fallen leaves, the Solomon Island eyelash frog (*Ceratobatrachus guentheri*), also known as the triangle frog, tries to go unnoticed throughout the day. Its pointed head and numerous small skin flaps help it pass for an unappetizing piece of dead vegetation. But at night this frog wants to be very much noticed by members of the opposite sex, and males call loudly from suitable patches of ground. Unlike most frogs who must return to water to breed, females of this species lay their clutches in shallow underground nests. There is no tadpole stage, and after about a month tiny but fully developed froglets hatch from the eggs.

The unusual looks and reproductive biology of this species make it a target of international pet trade. But while poorly regulated exportation from the Solomon Islands makes the eyelash frog the seventh most frequently exported animal from this country, it is severe habitat loss that may eventually spell the demise of this fortunately still common frog.

Warty tree frog *(Scinax boulengeri)* (Costa Rica)

Marbled tree frog *(Hyla ebraccatai)* (Costa Rica)

The striped tree frog (*Agalychnis calcarifer*) is a rarely seen species associated with natural forest gaps (Costa Rica).

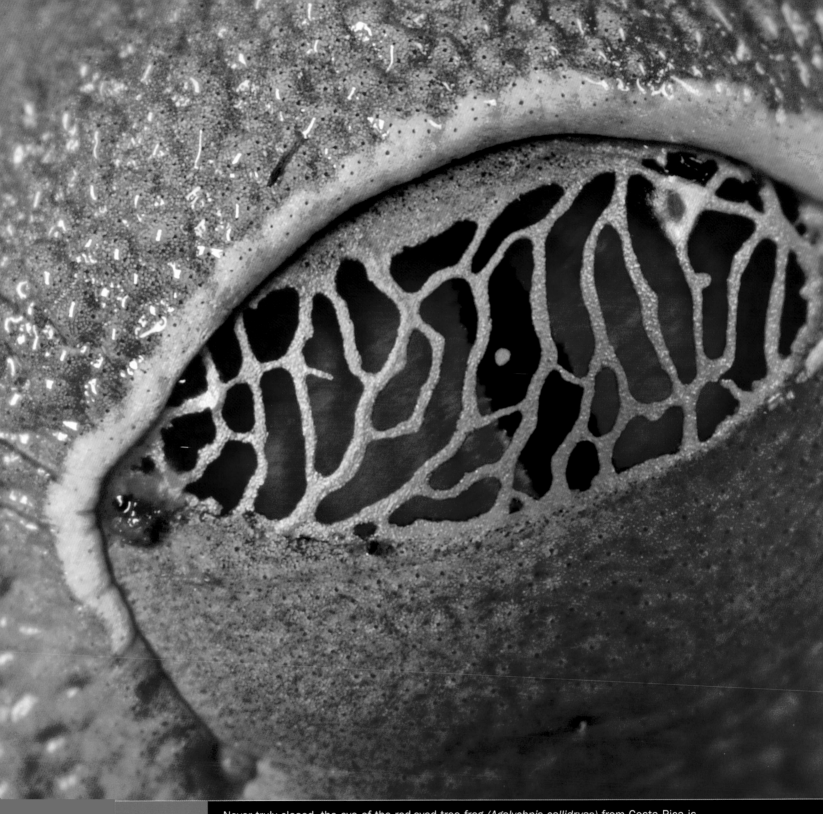

Never truly closed, the eye of the red-eyed tree frog (*Agalychnis callidryas*) from Costa Rica is

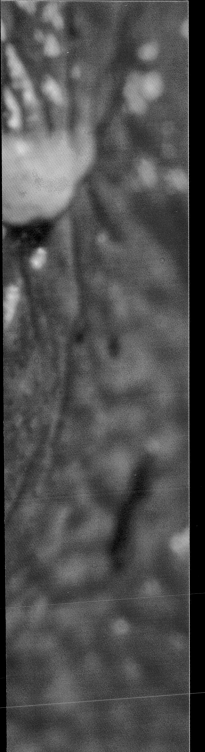

Malagasy web-footed frog *(Boophis luteus)* (Madagascar)

Red-webbed tree frog *(Hyla rufitela)* (Costa Rica)

The African golden frog (*Amnirana occidentalis*) is a very good jumper, difficult to approach. This species is only present in primary, undisturbed forests, making it vulnerable to habitat changes caused by logging and other human activity (Guinea).

The leopard running frog (*Kassina cochranae*) rarely jumps, preferring to run swiftly on its beautifully spotted legs (Guinea).

African tree frogs are not closely related to those in other parts of the world, and their adaptations to the arboreal lifestyle such as enlarged toe pads and grasping fingers are examples of evolutionary convergence. This *Leptopelis hyloides* is common in West African forests.

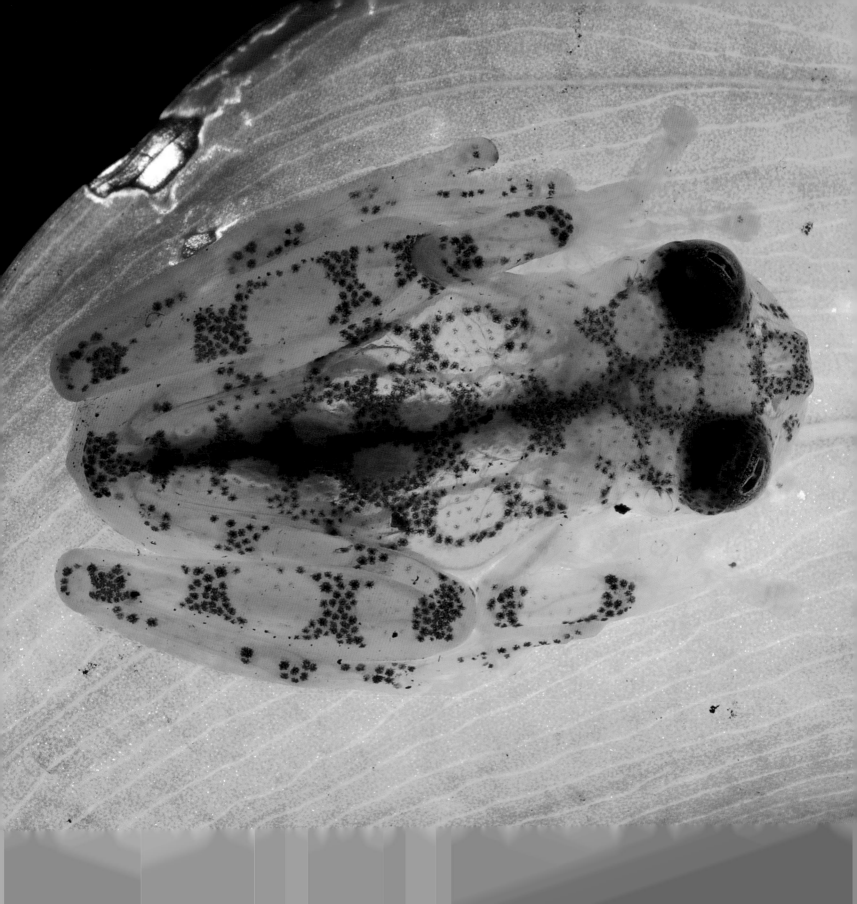

There are other ways to hide than crawling into tight spaces—you can become transparent. Many rainforest organisms have evolved ways of letting the sun's rays shine right through their bodies, thus appearing one with the leaves on which they rest during the day. This technique is not restricted to small, lightly sclerotized insects—some frogs such as glass frogs of the family Centrolenidae virtually disappear during the day. Some katydids, while not exactly transparent, achieve a similar effect by flattening their wings against the surface of leaves, and completely eliminating the shadow around their bodies.

A nymph of an unidentified planthopper (Guinea)

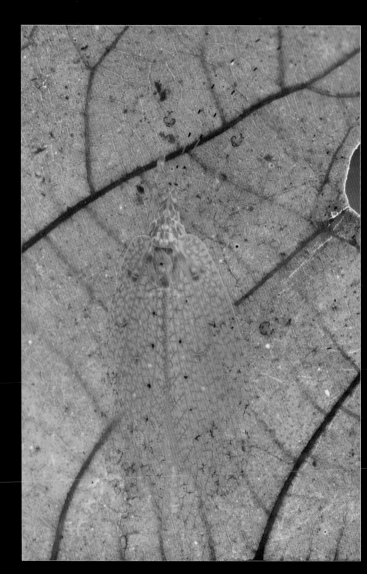

False leaf katydid *(Stenampyx annulicornis)* (Guinea)

61

Moss katydid *(Championica montana)*

Katydids

IF I WERE TO NAME ONE GROUP of organisms that encapsulates everything I love about nature, it would be katydids. They behave in intriguing ways, and their physical beauty is matched by their captivating songs. It also helps that katydids inhabit places I have always longed to visit. The splendor of katydids' coloration and the ornamentation of their bodies has long fascinated me. I cannot think of a more pleasing color than the subtly rich, lightly speckled green of a leaf-winged katydid. The intricate network of veins on their wings is undoubtedly one of nature's finest textures. Many are superbly adapted to

blend with their environment, in their crypsis giving us some of the best examples of natural selection at work. The illusion of foliage is often enhanced by lichen-like growths or faux "leaf damage." This masterful disguise fools not only their predators: I once got into a heated discussion with one of the best contemporary evolutionary biologists. He tried to convince me that the "lichen" on a young *Mimetica* katydid was real. In an attempt to prove his point he clipped a small piece of it and was stunned by the appearance of hemolymph, the insect equivalent of blood.

Yet not all katydids don a protective green coat. On the contrary, some species flash dazzling red, blue, or yellow markings, advertising their distastefulness or sharp armature. South American Crayola katydids (genus *Vestria*) always wear vivid blue and yellow coloration to ensure that monkeys and birds know what an unappetizing meal they would make. Other katydids such as the Australian mountain katydid *(Acripeza reticulata)* cover their blue and red abdomen with gray, leaf-like wings, revealing their warning coloration only when immediately threatened. The same tactic is employed by many moss and lichen mimics, who often flash brightly colored legs or abdomen to indicate that the plant-like ornamentation of their bodies is in fact a hard, spiny armor, capable of inflicting serious wounds. A swift kick from the moss katydid *(Championica)* drew blood from my finger, and I can only imagine what damage it can cause if applied to a bat's head.

Katydids have good reasons to use whatever protection is available. Birds, lizards, monkeys, and bats consider them a very tasty dish; in fact a study conducted on Barro Colorado Island in Panama has shown that katydids constitute about 60 percent of the food consumed by bats. Not surprisingly, their roosting sites can be easily identified by a thick carpet of katydid wings covering the floor. In Peru, katydids are the main component of the diet of tamarin monkeys, with different species of tamarins specializing in hunting different groups of katydids. But tamarins are not the only primates who value katydids as a nutritious meal, and across the globe humans of many cultures have eaten these insects. Native Americans used to feast on Mormon crickets (despite the misleading name, Mormon crickets are katydids), and many African tribes still eat some larger species. A researcher at the National Institute of Biodiversity (INBio) in Costa Rica is considering commercial breeding of large *Steirodon* katydids after discovering how incredibly tasty they are. And it is not only nutritional value people are after—in China certain species of katydids were once considered a powerful aphrodisiac.

Perhaps it is not surprising that katydids were chosen for this purpose, since their reproductive behavior is truly remarkable. It begins, as it often does in the animal kingdom, with a song. Males of most species possess sound-producing organs on their wings, and use them both to attract females and to stake claims to prime territories or singing perches. Katydid songs, known as stridulations, are unique to each species and, as in birds, they can be used by an experienced naturalist to identify local faunas without the need to actually see any animals. Songs of different species vary in their duration, intensity, and pitch. Many are produced in frequencies too high for a human ear to discern, and naturalists need a specialized ultrasound detector to record them. But in others the call is clearly audible and may take the form of a simple buzz, individual chirps, or a melodious, bird-like trill. For centuries, nature aficionados in China and Japan have appreciated musical calls of katydids, and even now one can buy live katydids in miniature cages there to enjoy their love songs at home.

Although males are responsible for most vocal performances, females are not completely silent. They respond to the male with short clicks, expressing their interest in his courtship and helping him navigate tangled vegetation. Unfortunately, females are not the only listeners to the male katydid's serenades. Bats, parasitic flies, and even geckos use these calls to locate their prey. For this reason some katydids have evolved means of communication that do not rely on aerial broadcast but use quieter, less detectable modes.

Once the female is appropriately charmed, she approaches her suitor. But now it is his turn to be coy. Unlike most animals, male katydids invest much more in their offspring than just a small dose of sperm cells. During mating they produce a large, nutritious packet of proteins, one that the female gladly devours and uses to produce her eggs. This packet, known as the spermatophore, can weigh as much as 40 percent of the male's body. This severely limits the number of times a male can mate and makes his choice of a partner quite critical. It is not uncommon for a male katydid to reject his partner if he judges her too small to be worthy of his investment. Under extreme conditions, when water and food are scarce, the katydid courtship undergoes an amazing role reversal, and females start to compete aggressively for males, a phenomenon exceptionally rare among animals.

Following an elaborate courtship, females begin to lay eggs. Some use their sword-like ovipositor to make incisions in plant stems and hide eggs there, others lay them in clusters on leaves, while still others bury them deep into the ground. Emerging young katydids usually resemble minute versions of their parents, but in certain species young nymphs are superb mimics of ants or wasps, gaining protection from most predators. The disguise, however, does not trick their katydid relatives. Many are voracious hunters, feeding on other katydids, caterpillars, snails, or even small lizards. Yet the majority of species includes at least some plant material in their diet. Powerful mandibles of conehead katydids easily crush even the hardest seeds, whereas leaf-winged katydids prefer softer parts of plants. An unusual adaptation for feeding on nectar and pollen of flowers is found in Australian pollen katydids.

Katydids, like most groups of insects, flourish in humid, tropical regions of the world. The highest diversity of these animals can be found in evergreen forests of Central and South America, Africa, and Southeast Asia. A single site in Costa Rica can harbor up to 150 species of katydids, while at a location in Peru researchers found over 300 species. There are about 6,800 known forms of katydids, but many more are still remaining to be discovered. We know little about species living in the forests of equatorial Africa, and a recent study of the Oriental region revealed that over 40 percent of katydids found there were new to science. But their diversity does not shield them from a fate sadly shared with many other organisms. Their habitats are shrinking or disappearing, and industrial pollution pushes them out of areas where they were once common. In fact, katydids have the dubious honor of being one of the few groups of invertebrates in which documented cases of species extinction have been recorded. The California shield-backed katydid from Antioch dunes was named *Neduba extincta* because it had disappeared forever shortly before being recognized by scientists, based on a few preserved specimens in a museum, as a species new to science. A similar scenario will unquestionably replay itself many times in the future, since with each patch of tropical forest lost to logging, we are also losing a unique habitat for a number of katydid species.

There are faint rays of hope for at least some katydids. In Europe several species have been taken under protection, and a captive breeding program has been initiated for a few particularly threatened ones. The situation is far more critical in the tropics, where nothing short of a complete reversal of current practices of indiscriminate clearcutting and forest burning will save countless species of katydids and other organisms. To make matters worse, our knowledge of what lives in some of the most threatened tropical habitats is fragmentary, making legal actions against environmental offenders more difficult.

It does not help that most katydids do not disperse well and have very limited ranges, often restricted to a single valley, mountaintop, or island. One night during a recent trip to the Solomon Islands, an archipelago in the South Pacific, I spotted a mysterious little animal. After a few seconds of an intense search in my mental database of katydid taxonomy, I realized that I was looking at a completely new species, most likely belonging to the Pacific genus *Ocica*. Holding my breath, I slowly reached for it. Alas, its long antennae detected my movement and with a powerful jump it disappeared in a dense thicket of ferns. I looked for it, unsuccessfully, for a very long time. On my way back to the village where my friend and I were staying I passed a large area of felled trees and noticed that since the previous evening a few more logs were lying on the ground. At that moment I realized that I might be the only human being who would ever see this unique, mysterious life form.

Once considered the rarest of the rare, the giant leaf katydid *(Celidophylla albimacula)* is actually quite common in lowland rainforests of Nicaragua and Costa Rica. As more researchers and naturalists enter tropical forests at night, armed with portable lights, infrared cameras, and ultrasound detectors, we begin to realize that many animal species previously known from single, opportunistically collected individuals are in fact fairly abundant, yet living in spatial or temporal realms that are quite different from ours. Rarity is a concept that should be applied to tropical invertebrates carefully. More often than not it simply reflects our lack of understanding of the animal's habits and seasonal or diurnal movement patterns. Most rainforest katydids are active only at night, safe from birds' sharp beaks and monkeys' agile hands. Days are spent resting in rolled-up leaves, under tree bark, or pressed tightly to branches and stems. Species that mimic leaves rest fully exposed during the day, and yet their superb camouflage makes them virtually invisible among the forest's vegetation.

Rhinoceros spearbearers *(Copiphora rhinoceros)* are voracious predators. They have been known to eat insects, snails, or even small lizards. Their activity starts at dusk and continues long into the night as they patrol vegetation, searching for sleeping animals, fruits, or seeds. Here a female captured a leaf-winged katydid *(Orophus tessellatus)* (Costa Rica).

A head armed with a prickly cone provides good protection against foliage-gleaning bats and monkeys, the principal enemies of the spearbearers, ferocious katydids of the Central and South American genus *Copiphora*. In forests full of animals looking for a katydid meal, these katydids equipped with strong, defensive armature are not afraid of eavesdropping predators and produce very loud, continuous calls. The differently shaped cone and facial color patterns of each species may indicate that these features play a role in species recognition. The powerful mandibles of spearbearers easily crush plant seeds and shells of snails, and some species are known to attack even small lizards. Their legs are armed with sharp spines, which help capture insects and other prey. Spiny legs also provide powerful defensive weapons, and my fingers have bled many times after handling these little warriors.

Yellow-faced spearbearer *(Copiphora cultricornis)* (Costa Rica)

Brown-faced spearbearers *(Copiphora hastata)* (Costa Rica)

A newly discovered, yet unnamed katydid (*Ischyra* sp.) in flight (Costa Rica)

Most leaf-winged katydids (Phaneropterinae) are excellent fliers. In Costa Rica the uppermost layers of the rainforest are home to the greatest diversity of forms, not surprisingly all superb leaf mimics. Many species never descend to the ground and complete their entire developmental cycle high in the canopy. Unlike most other katydids who deposit their eggs in the soil or in tree bark, theirs are laid directly on leaves, or are carefully inserted between layers of the leaf epidermis using a highly modified flat ovipositor.

Katydid *Ectemna dumicola* during its final molt (Costa Rica)

Leaf-footed katydid (*Aegimia* sp.)

The black, blue, and orange coloration of *Plastocorypha vandicana* probably serves a dual function—its rust-colored wings help it blend in among dry leaves, but if found by a bird the conspicuous facial markings will help the predator remember a painful bite from the mandibles of this seed feeder (Guinea).

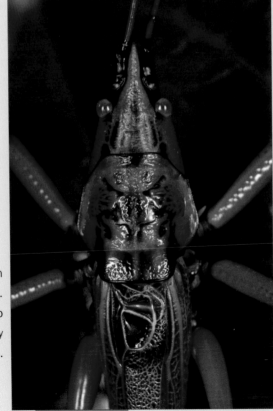

Bright coloration of *Odontolakis* sp. advertises a sharp armature of its body (Madagascar).

A young nymph of *Zeuneria melanopeza* looks nothing like its green, leaf-like parents (Guinea).

Not all katydids are green and leaf-like. Many species advertise the distastefulness of their bodies or the strength of their mandibles with gaudy colors, while in others bright, conspicuous markings imitate fungal diseases. Young nymphs, like West African *Zeuneria melanopeza,* frequently look very different from their parents, and taking advantage of their minute size, they mimic ants, small wasps, or spiders. Their behavior is also different from that of the adults— instead of slow, deliberate motions typical of adult katydids, they move in a fast, jerky fashion, flicking their antennae in a way eerily reminiscent of stinging wasps.

In the case of Costa Rican leaf-wing *Euceraia insignis,* the yellow, red, and blue markings help it blend among partially decayed leaves, rather than advertise anything; this species does not produce repellent compounds and, if found, is gladly consumed by birds and other predators.

A brown, dead-leaf form of
Mimetica crenulata (Costa Rica)

A fallen, partially decomposed-
leaf form of *Mimetica viridis*
(Costa Rica)

It would be difficult to find a better example in the animal kingdom of plant mimicry than the walking leaves of the genus *Mimetica*. These Central American insects are such superb masters of camouflage that locating one, especially during the day when they remain completely motionless, is next to impossible. At night, when they are feeding on leaves, it is a little easier to find them because the movement of their long antennae reflects the light of a flashlight in the still air of the rainforest and betrays their animal provenance. The details of their disguise are exquisite, down to a clear window and irregular incisions on their wings that look as if something has taken a bite. Young nymphs of *Mimetica* and related katydids, lacking wings and thus the ability to resemble leaves convincingly, develop lichen-mimicking growths on their abdomen, often with a masterful touch of fake mold. Coloration of walking leaves is extremely variable, and frequently individuals within the same population look as if they belonged to several different species. Such variability prevents visually oriented predators, primarily birds and monkeys, from developing a search pattern for these otherwise vulnerable insects. The inconsistency of the walking leaves' appearance also plays tricks on entomologists, and many species in this group have been described repeatedly under different names by researchers who failed to see consistent morphological traits behind their masterful costumes.

Attacks by certain moths or beetles cause leaves to turn brown on only a portion of their surface; this *Mimetica mortuifolia* makes a convincing impression of such a leaf (Costa Rica).

The ears of katydids are located on their front legs, just below the knees. They are finely attuned to calls of conspecific individuals and allow these animals to locate the caller with extreme precision. The brown ear membranes can be seen on the giant helmeted katydid *(Phyllophorella woodfordi),* although, interestingly, this particular species does not sing. Its simple ears are probably used only to detect the approach of potential predators, especially bats. But in some forms of katydids the hearing organs develop into rather sophisticated structures, in their shape sometimes resembling external ears of mammals.

The many islands of the Australasian region are home to some of the world's most unusual katydid forms. The emerald and red pandanus katydid *(Paramossula basalis)* is found only on the Solomon Islands, whereas spiny *Phricta spinosa* occurs in the tropical forests of northern Queensland. The giant helmeted katydids from New Guinea, the Solomon Islands, and northernmost Australia are some of the largest insects on the planet, with one of the species possessing a wingspan of thirty centimeters—nearly a foot. In an unusual, beetle-like adaptation, the neck region of the body, or the pronotum, of *Phyllophorella woodfordi* from the Solomon Islands develops into a large and hard shield that covers part of the abdomen, both protecting it and helping the animal blend into the green background of the forest.

Paramossula basalis

Phricta spinosa

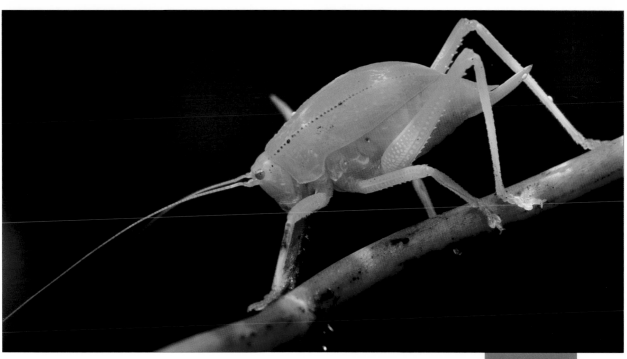

Phyllophorella woodfordi

Normally hidden *(above)*, but here photographed on a piece of glass *(left)*, *Haemodiasma tessellata* reveals striking warning coloration and heavy armature of the legs (Costa Rica).

One of the best moss mimics of the katydid world, *Championica montana* (Costa Rica)

Few katydids feed on moss, but many take advantage of its ubiquity in the rainforest, and have evolved coloration and body ornamentation that allow them to blend seamlessly into its thick carpet. Unlike leaf-mimicking katydids that often achieve complete crypsis with coloration alone, moss-mimicking forms had to adapt nearly every element of their bodies to create the illusion of fragmented, small-leaved vegetation. Thick spines on their legs and the neck region (pronotum) not only help mask the outline of the insect's body, but also provide excellent defensive weaponry. If detected by a predator, most species viciously kick their legs, fan their strikingly colored wings, and raise their abdomen to reveal bright coloration that warns of severe consequences to anyone trying to swallow the seemingly defenseless animal. Their armature combined with virtual invisibility protects moss katydids from eavesdropping predators, and many produce loud, long calls.

A particularly rich fauna of moss-mimicking katydids can be found in forests at higher elevations in the tropics, where nearly constant mist and slightly lower temperatures create perfect conditions for growth of moss and lichens.

Under cover of nearly complete darkness, illuminated only by a faint glow of bioluminescent mushrooms and rays of moonlight filtered down through millions of over-lapping leaves, a female Aztec katydid (Stilpnochlora azteca) clears with her mouthparts a patch on tree bark to lay a batch of eggs. It is a common pattern among insects and other smaller animals that any activity making them more vulnerable to predators happens at night. Thus katydids, like many other insects, feed, court, mate, and lay eggs well after sunset. The few that prefer to be active during the day hide in impenetrable thickets of grass or high in the treetops where few pred-ators venture. Nights also tend to be more humid and less windy than days, and therefore more suitable for activities that may involve serious water loss or cause bodily harm if interrupted by a sudden gust of wind, such as hatching from eggs or molting.

For this Aztec katydid and her offspring, the struggle to stay alive begins early. The eggs she has just laid will probably hatch in a month or two if they are not found first by a wasp of the family Encyrtidae. These tiny wasps belong to a group of insects known as parasitoids. Unlike parasites, which tend to keep their host organisms alive, parasitoids invariably kill their host. In this case, the wasp larvae develop inside the katydid eggs, slowly but surely devouring the embryo. It is not uncommon for the entire batch of katydid eggs to produce only wasps and not a single katydid. However, the genetic code in this female katydid has a contingency plan, and she will lay several additional batches of eggs, at different times and in different locations. Even if only one of her nymphs survives to adulthood, she will achieve her goal of passing on her and her partner's genes.

The Aztec katydid belongs to a group that lays eggs on the surface of leaves, bark, or twigs. This is typical of forest canopy species, many of which never descend to the ground. Their eggs are very hard and resistant to desiccation, and they can sit exposed to the elements for weeks, losing little water. The advantage of this strategy, which offsets being an easy target to wasps and other predators, is that newly hatched nymphs find themselves in the right habitat and can begin feeding immediately. Other species come down from the trees to lay eggs in the ground, where they remain safe from parasitic wasps and water loss until hatching. But then the young, unarmed nymphs must negotiate their way back to the treetops, a perilous journey studded with ambushes from spiders, marauding ants, and hungry lizards. In the end, both strategies tend to ensure the survival of just enough individuals to carry on with the reproductive cycle.

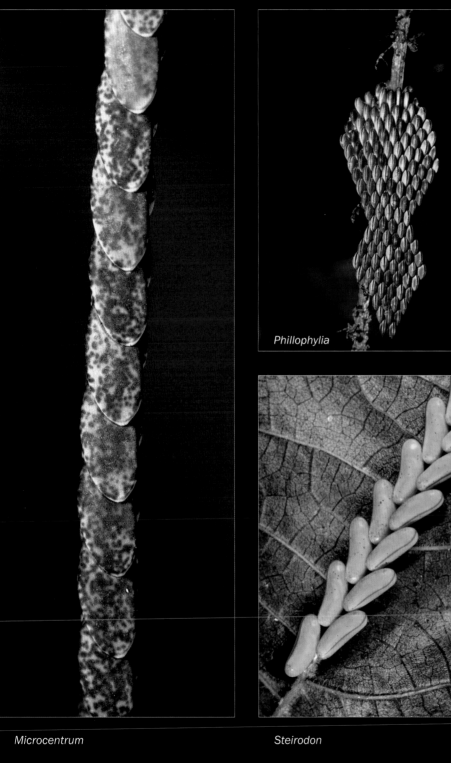

Microcentrum

Phillophylia

Steirodon

Unknown katydid

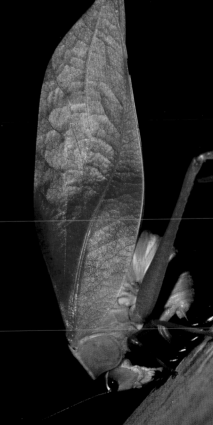

0:00

0:38

0:41

Unlike vertebrates, which support their bodies with an internal skeleton hidden within soft tissues, insects and other arthropods protect their internal organs and support the muscles with a hard, external armor, the cuticle. This external skeleton achieves its rigidity thanks to the presence of chitin, a polysaccharide that forms long, straight microfibers embedded within a protein matrix. The surface of the cuticle is covered with a thin layer of waxes that provide the animal with water impermeability. Such a design of skeletal structures appears to be superior to the internal one, as evidenced by the unparalleled success of arthropods, which currently constitute the majority of animal life on Earth.

But there is no such a thing as a perfect design, and the arthropods' Achilles' heel is the need to periodically shed their exoskeleton to allow for the growth of the body. The new cuticle forms under the old layer and

separates very shortly before molting. Molting, or ecdysis, is a process that leaves the animal utterly vulnerable. As soon as the new cuticle hardens enough, the arthropod takes a few halting steps to hide until its new cuticle is fully formed. Molting requires just the right humidity and temperature. The majority of insects undergo molting at night, when the risk of being noticed is lower, and the humidity is higher.

Most insects take advantage of gravitational forces to help them exit the old cuticle, and molt hanging upside down from a branch or other high point above the ground. But some such as cockroaches (opposite, below left), certain crickets, or spittle bugs (Cercopidae) are capable of molting on horizontal surfaces. Many insects with chewing mouthparts eat their old cuticle after molting, recovering all proteins and carbohydrates lost during the process.

0:53

1:07

1:49

The entire process of molting in a Costa Rican katydid *(Xestoptera cornea)* takes about two hours. Once its wings are fully formed and the exoskeleton is hardened enough to allow walking, the insect consumes its old cuticle and spends the remainder of the night hiding and allowing its body to fully recover.

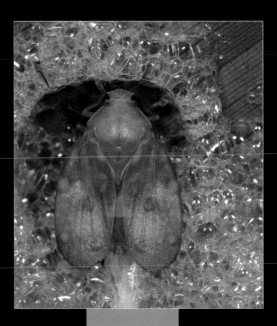

A spittle bug (*Tomaspis* sp.) undergoes its final molt in the foamy froth in which it has spent its entire larval development (Costa Rica).

Whereas nocturnal soundscapes are created primarily by crickets and katydids, daytime belongs to cicadas. Almost any insect sound heard during the day in tropical forests is produced by a pair of large, vibrating membranes underneath the abdomen of a male cicada. Few singing animals, including most birds and frogs, can compete with the force of a cicada's love song. In some cases the sound produced by a single male can approach 110 decibels, a level comparable to that of a chain saw. Unfortunately, their calls are also far less pleasing to the human ear than those of birds, and many resemble heavy machinery in dire need of lubrication. It is quite remarkable that in the deafening chorus of thousands of males a single wing click produced by a female when she finds an attractive suitor can be heard by him.

Most of a cicada's life, however, is spent silently underground, where the nymph, equipped with a pair of powerful digging legs, sucks juices (phloem) out of roots. Some species spend many years, sometimes more than ten, in the soil before climbing out and undergoing their final molt. Empty "shells" (exuvia) of cicadas can often be seen on tree bark in tropical forests, especially after the rainy nights that many of them choose for molting. Their adult life is short by comparison, lasting no more than several weeks. Most of it takes place high in the canopy, where their excellent camouflage, good vision, and powerful wing muscles allow them to evade larger predators. But like other singing insects, cicadas often fall victim to parasitic flies, perhaps led to them by their song.

A male emerald cicada
(*Zamara smaragdina*)
completes its final molt;
a shield-like structure
behind its hind legs
covers thick mem-
branes used to
produce his loud
call (Costa Rica).

One night in Costa Rica I noticed a rather curious sight: a large cockroach was sitting on top of a smaller lantern bug, or a fulgorid, a distant relative of cicadas. The fulgorid did not seem to mind and continued to feed on tree juices, while the roach appeared to be licking something off the fulgorid's wings. Anxious to find an explanation for this behavior, I spent many nights watching these animals and soon realized that several species of moths and ants also take advantage of the fulgorids. Fulgorids, like most members of the insect order Auchenorhyncha, feed on plant juices (phloem), which have very low protein content but are rich in sugars. The excess water and sugar is regularly expelled in the form of sweet droplets, or honeydew, that other insects eagerly seek. Ants are well known to feed on honeydew produced by aphids, but this type of trophic relationship had not been seen in much larger fulgorids. In exchange for their sugary offering, ants award aphids protection from predators. It is unclear if the fulgorids benefit in any way from the presence of moths and cockroaches. Perhaps they simply tolerate these sugar-seeking insects without any reciprocal benefit because they must relieve themselves of the extra water and sugar anyway, and trying to evade the freeloaders could be costly or even dangerous. It is also possible that the presence of a conspicuous insect may distract a potential predator from a harmless fulgorid.

Moths of *Platynota* sp. (Tortricidae) and an unidentified species of a noctuid (Noctuidae) compete for access to honeydew produced by a fulgorid *(Enchophora sanguinea)*. Fulgorids are very host-specific, and the same individual stays on its host tree for many nights, providing a reliable source of sugar for a variety of animals (Costa Rica).

Large cockroaches
of the genus *Eurycotis*
take turns at the tip
of the abdomen of a
fulgorid, *E. rosacea*
(Costa Rica).

A cluster of small, short-lived flowers of *Hamelia patens,* a relative of the coffee plant, is the stage for a surprisingly intricate collection of biological relationships. This plant depends on hummingbirds for its pollination, and lures them with its bright red flower corollas and copious quantities of nectar. But in addition to the most welcome pollen from other plants, the birds unknowingly bring tiny mites, *Proctolaelaps kirmsei,* which hide in their nostrils. Despite their minute size, which barely reaches one millimeter, these mites effectively compete with the birds for nectar. The mites' appetite forces hummingbirds to spend two extra hours visiting flowers each day to gather the same amount of nectar the birds would have collected had the mites been absent. After just a few days *Proctolaelaps* mites, sensing the approaching death of their flower cluster, or an inflorescence, hop on the next visiting bird to move to a fresh one.

But hummingbirds and mites are not the only animals interested in the sweet nectar of *Hamelia.* Large ants *Ectatomma tuberculatum* jealously guard the inflorescence and gather nectar in their large, scoop-shaped mandibles. They do not hesitate to use these mandibles to chase away, and if necessary kill, any intruders. Their ferocity makes them a good model to mimic, which has been achieved by this otherwise harmless true bug *Hyalymenus* sp. (Alydidae).

Ant-mimicking true bug (*Hyalymenus* sp.)

Ant (*Ectatomma tuberculatum*)

Flower mite
(*Proctolaelaps kirmsei*)

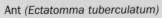

89

Despite its widespread use, the word "bug" should only be applied to one particular group of insects, the order Heteroptera. These "true bugs" include a large number of families of insects characterized by sucking mouthparts and the unique structure of their front wings (hemielytra), which are divided into hardened anterior and membranous posterior parts. These wings are functionally similar to the tough, armor-like forewings (elytra) of beetles, and many true bugs can be easily confused with these insects.

True bugs feed on a wide variety of plant and animal material. Predaceous forms feed in a way similar to spiders—after perforating their prey with a sharp proboscis, they inject it with digestive enzymes, which liquefy the soft tissue of the victim and allow the bug to drink it. A number of species of true bugs are parasitic, feeding on blood of mammals and other animals. The ultimate bug, the human bed bug (*Cimex lectularius*) also belongs to this group. But the greatest majority of true bugs derive their nutrients from plant tissues and sap.

A high proportion of true bugs show bright aposematic (warning) coloration. In most cases their color pattern advertises chemical protection provided by abdominal or thoracic glands, which produce pungent and very effective repellents. Stink bugs (Pentatomidae) *(above)* are named after the characteristic, sharp smell they produce if attacked by a predator.

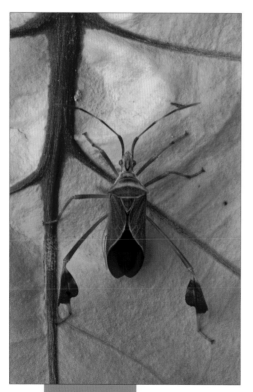

Leaf-footed bug (Coreidae)
(Dominican Republic)

A shield bug nymph (Scutellaridae) (Guinea)

The thorn bug *(Umbonia crassicornis)* (Costa Rica)

Nymphs of treehoppers can look very different from adults,
as illustrated by this Costa Rican treehopper *(Membracis dorsata).*

Treehoppers (Membracidae) are relatives of cicadas and lantern bugs. Much like these insects, treehoppers feed on juices of plants, sucking on them with their long proboscises. The bodies of young tree-hoppers are frequently adorned with long projections that probably protect these insects from at least some predators. The adult treehopper covers almost its entire body with a shield of the enlarged pronotum, a cuticular plate situated right behind the head of most insects. The pronotum frequently assumes the shape of thorns, similar to those of plants on which these insects feed.

Many treehopper species benefit from a symbiotic relationship with ants, who defend them from predators in exchange for honeydew produced by the treehoppers.

Tarantula *(Psalmopoeus reduncus)* (Costa Rica)

Arachnids

THERE IS SOMETHING SPECIAL ABOUT arachnids. Spiders and scorpions and, to a lesser extent, mites, evoke exceptionally strong emotions from people—negative as a rule, but sometimes also bordering on fanatic obsession. I have yet to meet a person who has no opinion about arachnids. What aspect of their appearance or biology makes such a remarkable impression? Surprisingly, the venomousness of some of these animals is rarely given as the reason for the dislike. My sister, for example, claims that she is afraid of spiders because they have "too many legs" (although she has no

problem with millipedes), whereas in my wife's case it is their "unpredictable agility." On the other hand, a friend of mine was able to survive a bitter divorce thanks to the solace he found in his love of mites and the study of their diversity. Tarantulas and scorpions almost rival aquarium fish and parrots in their popularity as pets, and many people consider an orb web to be one of nature's most beautiful designs. Regardless of what we think of arachnids, there is no escaping them. Our own bodies are home to several species of mites, and wherever you live you are always within a few feet of a spider.

Arachnids are some of the most successful life forms on our planet. With the exception of the deep waters of oceans and lakes, these animals have colonized virtually every available habitat on Earth. There are approximately 97,000 known species of these animals, but this number represents only a fraction of the real number of species, most of which remain unknown to science. A factor that played a very important role in their ubiquity and high species richness is the exceptional ability with which many arachnids can disperse and populate new areas. Unlike insects, arachnids have never evolved wings, and thus powered, fully controlled flight. And yet many mites and spiders move through the air with surprising ease. Mites are generally tiny and thus very light-bodied. Most species do not exceed one millimeter, and their immature stages are even smaller. Such small organisms do not require wings to be airlifted, and some species travel across and between continents in air currents, sometimes reaching the higher strata of the atmosphere. Thus, mites often colonize even the most remote terrestrial habitats, such as tiny oceanic islands and high mountain

peaks. Once settled in a new area, and having be-come genetically isolated from other populations of their own species, mites may undergo an evolu-tionary process known as allopatric speciation. Isolated populations of organisms are often subject to different selective pressures than those of their ances-tral populations (for example, they may need to adapt to feeding on different species of plants or to survive in slightly higher temperatures), resulting in an accu-mulation of genetic changes that eventually lead to the appearance of a new, distinct species. On the other hand, mites' boundless peregrinations may sometimes have quite the opposite effect, and some species have a large, almost cosmopolitan distribu-tion. Naturally, in addition to the passive flight, many free-living mite species are transported with soil, plants, or, in a behavior known as phoresy, as unin-vited passengers on bodies of other animals.

Some spiders, being generally larger than mites and thus having smaller body surface area relative to their weight, have evolved a passive flight mechanism known as "ballooning." Young individuals of many spider species spin a single strand of silk, which effec-tively increases their surface area and allows them to be airlifted and moved to a new location. Mass migra-tion of young spiders is a part of an annual phenom-enon known as Indian summer.

But their ability to disperse over large areas is not the only reason for the arachnids' success. They ex-hibit a range of astonishing behavioral and physiolog-ical adaptations that allow them to survive in some of Earth's harshest environments. Scorpions can survive in deserts in part because of their ability to reflect ul-traviolet light, thereby minimizing the negative effect of excessive solar radiation. In the same habitat, mites survive because they are able to live deep in the sand, protected from both high temperatures and extremely low humidity. Certain phoretic species of mites that attach themselves to the bodies of insects may enter a period of dormancy to survive the trip. During this pe-riod their mouthparts become nonfunctional, and in order to avoid death through desiccation they actually drink through their anus! On the other end of the environmental spectrum, some spiders are able to spend their lives under water thanks to their ability to build chambers filled with air, allowing them to breathe. Spiders in the temperate zones are remarkably tolerant of low temperatures. Some can survive being completely frozen in a block of ice because of the presence of antifreeze proteins in their bodies.

Another reason for the success of some arachnids is highly developed maternal care, which significantly increases the chances of survival of the young. Spiders build strong, protective egg sacs made of silk that protect the eggs from desiccation and shield them from predators. Female scorpions, tailless whip scorpions (Amblypygi), vinegaroons (Uropygi), and many spiders not only protect their eggs but also carry newly hatched young on their backs until the juveniles are able to fend for themselves.

With the exception of several lineages of mites, all arachnids are predators. Some groups, namely spiders, scorpions, and false scorpions (Pseudoscorpiones), have independently evolved venom glands, each with a different delivery mechanism for the toxic compounds. Spiders inject their venom into the body of their victims with a pair of fang-like mouthparts known as the chelicerae. Scorpions have a venom gland and a single stinger at the tip of their long abdomen, whereas false scorpions have their venom glands in their "pincers," or the pedipalps. The venom's function is not only to subdue or kill the prey, but also to help digest it. Spiders are incapable of chewing, so the only way they are able to get nutrients from the prey's body is to suck them out. Thus, their venom liquefies the tissues of their victims, allowing spiders to literally drink their meals. Because venomous arachnids feed almost exclusively on insects and other arthropods, their toxins have evolved to have maximum effect only on these animals. Only a handful of species possess venom that can cause death in humans.

As with other small animals, the main threat to the survival of many arachnid species is habitat loss. But in a few cases an unregulated pet trade has caused sharp declines in the populations of species popular among hobbyists. Seventeen species of tarantulas have been overcollected and are now protected. The giant emperor scorpions (*Pandinus*) of Africa are also considered vulnerable, and their trade is regulated by the Convention on International Trade in Endangered Species (CITES). These animals are particularly susceptible to overharvesting because of their long gestation period (at least seven months), small brood sizes (thirty to thirty-five juveniles), and length of time necessary to reach sexual maturity (four to seven years). A number of spider and false scorpion species, especially those associated with fragile ecosystems, such as caves, are considered vulnerable or endangered, and are included on the IUCN Red List of endangered organisms.

All scorpions are predators, capable of subduing even very large prey with their venom. Scorpion venom includes a complex mixture of peptides that damage the victim's nervous system by blocking potassium and sodium channels in its cells. Most of the components of the venom, such as the newly discovered peptide ardiscretin, have evolved to be especially effective on insects, scorpions' primary diet. They are generally not lethal to large vertebrates, such as humans. Smaller lizards sometimes fall prey to scorpions, but certain species of geckos show a resistance to their venom and hunt scorpions with impunity.

This conehead katydid (Copiphora hastata) had no chance against a large Costa Rican scorpion Centruroides limbatus, an arboreal species common on the Atlantic coast of this country. After encountering this drama in 1998, I actually stole the scorpion's meal. At that time the conehead katydid was new to science, and I needed all the individuals I could get to prepare its scientific description.

Newly born scorpions *Centruroides bicolor* stay on their mother's back, protected by her formidable stinger until they are capable of hunting on their own (Costa Rica).

The primary prey of the arboreal scorpion *Centruroides limbatus* are crickets and katydids (Costa Rica).

The UV reflectance increases as the scorpion's cuticle hardens; this individual probably molted recently and its UV reflectance is not yet fully developed.

One of the most amazing properties of scorpions' biology is the fact that their bodies reflect ultraviolet light (UV), causing them to glow when exposed to it. The adaptive value of the UV reflectance in scorpions is not entirely known. It is likely that it lessens the negative effects of excessive ultraviolet solar radiation in desert species. However, UV reflectance is present also in both cave dwelling and rainforest species, which are rarely, if ever, exposed to sun. Some researchers speculate that scorpions use differential fluorescence in different parts of their own cuticle to guide their movements toward light or dark areas, augmenting their rather poor vision.

The UV reflectance of scorpions is most likely caused by the presence in their cuticle of fluorescent beta-carbolines, although the precise nature of this phenomenon is still not fully understood. These substances and chemical processes that lead to the UV fluorescence are very similar to those found in human eye lenses affected by cataracts, a common cause of blindness. Understanding the mechanism that leads to scorpion fluorescence is now a focus of intensive research that may help develop new treatments for cataracts caused by oxidative protein aging in older people.

Scorpion *Centruroides limbatus* glows in ultraviolet light. The same individual looks very different when photographed in normal light *(opposite, above)*.

A pair of free-living velvet mites (*Trombidium* sp.) from Madagascar show significant sexual dimorphism, a common phenomenon among mites. The male is generally smaller, as in this case, but in some groups where male-to-male combat is frequent, the males are larger.

One of the smallest groups of arachnids is the Ricinulei *(opposite)*. This poorly known group of animals includes only fifty-seven described species in Africa and Latin America. Very little is known about their biology. They feed on small invertebrates, although the morphology of their mouthparts suggests that they may also be scavengers. Ricinulei are restricted to very humid habitats, such as caves or leaf litter and moss aggregations in the rainforest.

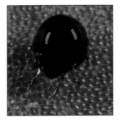

On the other end of the spectrum of arachnid diversity are mites. With nearly 39,000 known species, they are the most species-rich group of this class of animals. Still, researchers estimate that this number may represent a mere 5 percent of the actual number of mite species on our planet. They can be found virtually everywhere, even near the peak of Mt. Everest, and their biology and behavior are extremely diverse. Rainforest soil is particularly rich in mite species, but virtually every plant and many animals in the forest have their own, unique species of mites.

Phoretic mites are unwelcome passengers on a longhorn beetle, *Chlorida festiva* (Costa Rica).

The second pair of legs of males of ricinuleid
Ricinoides afzelii is enlarged and probably used in
combat with other members of its species (Guinea).

A characteristic "hood" covers mouthparts of all Ricinulei,
including this *Cryptocellus* sp. (Costa Rica).

Male *Agrioppe savignyi* (Costa Rica)

Female *A. savignyi*

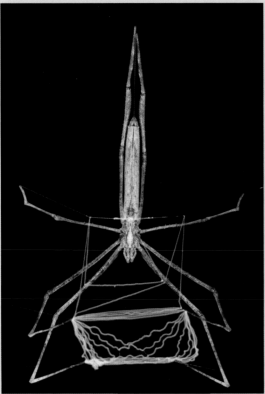

Spider silk is one of the strongest known organic materials, much stronger than steel of equal weight. It is also incredibly elastic, with individual strands breaking only if stretched two to four times their original length. The main component of spider silk is the polymerized protein fibroine, produced by glands opening in tube-like spinnerets at the tip of a spider's abdomen. At least seven different types of silk glands are known in spiders, each producing silk for a different purpose—walking threads, threads for encapsulation of the prey, silk for making an egg sac, and most important, sticky silk used by orb weavers to capture prey.

Orb webs of some spiders include thicker strands, which often form regular patterns. These structures are known as stabilamenta. Their function is not entirely clear. Some studies suggest that stabilamenta, which reflect UV light well, serve to redirect flying insects away from it and into the web. Others claim that stabilamenta are built mostly to conceal the spider from birds, which often pick spiders out of their webs.

Not all spiders build elaborate orb webs. Many actively hunt insects without using silk traps, while others produce irregular cobwebs. Phylogenetic studies reveal that irregular cobwebs are derived from the far more intricate orb webs, rather than the opposite. Some spiders, like the genus *Deinopis (left)*, do not wait passively for an insect to fall into their web. Instead, they cast the web onto it.

A female Costa Rican tiger spider *(Agrioppe savignyi)* entwines
a hapless firefly beetle (Lampyridae) that flew into her web.

A bark-jumping spider (subfamily Marpissinae) with a termite prey (Brazil)

Jumping spiders (Salticidae) are animals of great beauty and fascinating behavior. Their large eyes give them an almost mammalian quality, and not surprisingly these animals are visual, diurnal predators that stalk and hunt prey in a way reminiscent of cats. Eyes of jumping spiders can move in all directions, focus, and zoom on objects. It also helps that these spiders have four pairs of eyes, giving them a 360° field of vision. In addition, their thoraxes can turn 45° each way, allowing them to look around with even greater ease.

Given their superb vision, it is not surprising that jumping spiders rely on this sense in their mating behavior. Males perform elaborate, truly entertaining dances, flashing bright colors and performing species-specific moves that should convince the female of their worthiness as fathers.

Jumping spiders do not build webs but use silk as safety lines, anchoring themselves safely before making a long jump. And jump they can—many species can perform leaps at distances 50 times longer than their own body. Surprisingly, they do not have strong leg muscles but rely instead on sudden increases of blood pressure in their legs, which causes them to extend almost instantaneously, catapulting the spider into the air.

Anasaitis sp. (Dominican Republic)

Bathippus sp. (Solomon Islands)

Phiale formosa (Costa Rica)

Resembling living jewels, species
of the genus *Gasteracantha* are
unlike any other spiders. Their
abdomen, or opisthosoma, is
heavily sclerotized and armed
with hard spines. Although the
function of this armature is
unknown, it is possible that it
serves as a protection against
birds and lizards.

Gasteracantha metallica
(Solomon Islands)

Gasteracantha sp. (Solomon Islands)

Gasteracantha cancriformis (Dominican Republic)

Micrathena obtusispina
(Dominican Republic)

Eriauchenius sp. (Madagascar)

Poltys sp. (Australia)

Stephanopis sp. (Costa Rica)

Alcimosphenus licinus (Dominican Republic)

Sociality is not an attribute usually associated with arachnids. In fact, cannibalism is very common among these animals. But some members of a group known as false scorpions (Pseudoscorpiones) have evolved social behaviors that allow them to hunt prey much larger than any individual could overpower if working alone. False scorpions lack the stinger and "tail" of their much larger cousins, the true scorpions. Instead, their venom glands open at the tips of their "pincers," or pedipalps. The venom is not very fast acting, especially if the victim is large. The prey must be held in place long enough for the venom to act and paralyze it. Some species of the family Atemnidae form large family groups under the bark of rainforest trees, sticking out their grasping pedipalps to catch unsuspecting passersby. Many individuals, young and old, hold the prey, while at the same time injecting venom into its body, and soon the feeding can begin. Like spiders and other arachnids, false scorpions cannot chew and must inject their prey with digestive enzymes to drink the contents of its body.

When the colony grows too large, it is time to move on. False scorpions are master stowaways, and often attach themselves to bodies of flying insects in order to find new locations or mating partners.

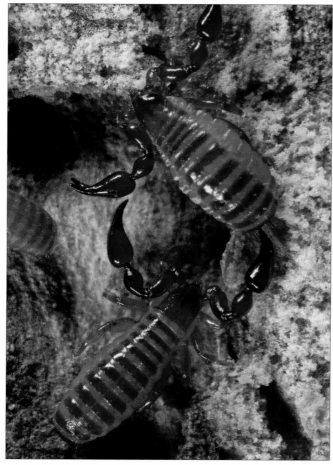

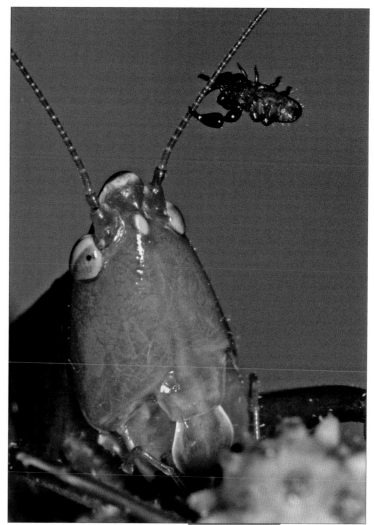

The body of a daddy longlegs resembles a gondola suspended on long, flexible stilts (Guinea).

Cynorta sp. (Costa Rica)

Some daddy longlegs are brightly colored, possibly to advertise their distastefulness (Guinea).

It seems like there is not much substance to a harvestman—all thin, spindly legs. But these arachnids on stilts, known also as daddy longlegs, are efficient predators that can move swiftly, both to capture prey and escape attacks. If necessary, they do not hesitate to reject one or more of their own legs, confusing predators and earning time to flee. Some species are gregarious and form large, constantly vibrating masses, from which a bird or another attacker is usually unable to extract a single individual.

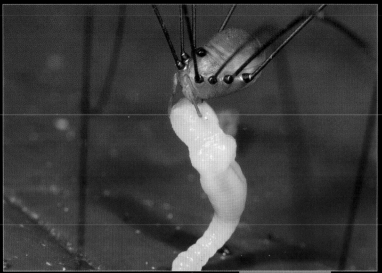

Daddy longlegs *Leiobunum* sp. devouring an arboreal earthworm (Costa Rica).

The whip scorpions, or amblypygids, are some of the fiercest-looking creatures of the tropical forests. Yet these distant relatives of spiders and other arachnids are entirely harmless to humans. Unlike spiders and true scorpions, they do not possess venom glands and must rely on the swift action of their spiny pedipalps to capture prey, such as crickets or other arthropods inhabiting tree trunks, the favorite habitat of many whip scorpions. Not being able to rely on their eyes, they detect their prey with their highly elongated, sensitive, antenna-like first pair of legs.

Whip scorpion males are very territorial and will engage in prolonged wrestling matches if they run into each other on the same tree trunk, especially if a female is nearby. Cannibalism is also frequent, and young individuals tend to avoid exposed surfaces on tree trunks occupied by older individuals.

A territorial combat of two male whip scorpions (Phrynus parvulus) on a tree trunk in the lowland Atlantic rainforest of Costa Rica.

Having just molted, this whip scorpion *(Phrynus parvulus)* will be vulnerable for a few hours until its cuticle hardens (Costa Rica).

Safe on their mother's back, young whip scorpions stay with her until they are ready to feed on their own (Costa Rica).

For a slow, seemingly harmless animal, millipedes (Diplopoda) have surprisingly few enemies. As is often the case among small animals, an innocent demeanor hides powerful and quite unexpected defensive strategies. Millipedes have two major lines of defense, both effective enough to deter virtually all predators. First, the chitinous base of their exoskeleton is composed of up to 70 percent calcium and magnesium carbonate, making it much harder than that of most terrestrial arthropods. Few predators are strong enough to crack it, although mongooses have been reported to toss ball millipedes *(Zootherium)* against rocks to break their shells. For smaller predators such as spiders, the hard exoskeleton is an almost impenetrable barrier. But if one tries its luck anyway, a rich arsenal of chemical warfare stops the predator in its tracks, literally. Some millipedes produce substances that act as sedatives, forcing the spider into a state of suspended animation that lasts for many hours. Interestingly, this substance is virtually identical to the drug quaalude, a synthetic sedative once widely used in medicine and as a recre-

ational, illicit drug. But licking millipedes to achieve any kind of a narcotic effect is not a good idea—many species instead produce cyanide, the most powerful inorganic poison known to man. Others produce substances that cause vomiting, headaches, and other unpleasant effects. And yet, few millipede species have warning coloration, which is typical of organisms that produce repellant chemicals. It is believed that the extreme effectiveness of their chemical weapons, combined with the very long time they have been around (terrestrial millipedes date back to the Carboniferous), made birds and other visual predators develop a genetically based aversion to these animals, and the millipede's shape alone is a sufficient warning.

The one enemy millipedes have no good defense against is, of course, human activity. Chemical pollution by "calcium analogues," strontium and fluoride, disrupts the formation of their exoskeleton, and desertification of many habitats affects species of millipedes that depend on high humidity and availability of moist, decaying leaves for food.

Male millipedes often guard their mates by riding on their backs (Costa Rica).

Millipedes of the genus *Platydesmus* are highly gregarious (Costa Rica).

Drops of repellant on the body of a millipede (Solomon Islands)

Pill millipede (*Zootherium* sp.) (Madagascar)

119

A giant centipede (*Scolopendra* sp.) (Solomon Islands)

The centipedes, unlike their slow, herbivorous cousins, the millipedes, are lean and mean predators. Powerful venom produced by glands at the base of their mandibles can overpower even small vertebrates, such as birds and mice. While no species of a centipede is known to be lethal to man, their bite is exceptionally painful and causes swelling and fever.

In an interesting example of a parallel evolution, some centipedes such as *Scutigera* have evolved compound eyes, similar to those found in insects.

A long-legged *Scutigera* cleans its foot (Costa Rica).

Ants (*Odontomachus hastatus*) attacking a cricket (*Aclodes* sp.) (Costa Rica)

Fungus-feeding cricket (*Phalangopsis* sp.) (Costa Rica)

Each night in the forests across the tropical belt millions of long-legged tree crickets (Phalangopsinae) emerge from hiding under bark and fallen tree trunks to court, mate, molt, eat, and be eaten. Many species of this group are gregarious, forming large assemblages between buttress roots of rainforest trees. This behavior most likely serves as a protection against predators, who will have difficulty picking out individual crickets from the crowd. This is the same strategy employed by some tropical daddy longlegs. The proximity of other members of their species also relieves the males from the need to loudly advertise for a mate. In fact, some of these crickets no longer rub their wings together to produce an audible song, but instead flick their wings in a vertical position, producing a strong puff of air that the females apparently find as arousing as a more traditional song. This close-range communication carries the additional benefit of not being detectable by eavesdropping predators. The female is able, at her leisure, to assess precisely the size and genetic quality of her suitor.

Tree crickets are so abundant on some trees that they constitute the principal diet of some spiders, scorpions, geckos, preying mantids, and ants. The crickets themselves feed mostly on plant material, including many species of fungi. They probably play a role in the dispersal of fungal spores, as the virtually still air of the rainforest interior provides little help with moving the spores away from the parental fruiting body.

Cricket (*Paragryllus* sp.) molting (Costa Rica)

Fungus-feeding cricket
(*Phalangopsis* sp.) (Costa Rica)

A leafcutter ant *(Sericomyrmex amabilis)* on a fungus garden (Costa Rica)

Soldier African driver ant (*Dorylus* sp.) (Guinea)

Ants

Barring a strong case of myopia, the animals we see most often are ants. This statement may come as a surprise because most of us have acquired the ability to filter them out of our field of vision, but the truth is that for every pigeon, butterfly, or snail, our eyes register probably between ten and one hundred ants. In temperate zones ants are some of the first animals to appear in the spring and the last to disappear before winter, while in the tropics it is sometimes hard to find an animal that is not an ant. Lean against a tree in the rainforest and within seconds you will find yourself at the center of attention of very curious and hungry ants. Leave a piece of organic matter on the ground and within minutes it will be swarming with ants. Even as I write these words, carpenter ants are chewing the walls of my house, causing expensive damage. And yet I cannot stop admiring these wonderful animals.

Ants are unquestionably some of the most successful life forms that ever existed on this planet, achieving diversity and abundance unparalleled by virtually any animal group, humans included. The key to their success is, of course, their sociality. With the exception of human societies, no other organism has created communities as extensive and finely tuned as those of ants. Ants belong to a category known as eusocial organisms, where several generations overlap and care for the offspring. Among invertebrates, eusociality is also known in bees, wasps, termites, ambrosia beetles, and some coral reef shrimp, but ant colonies overshadow those of most other animals in sheer size and complexity. Recent studies in Europe have uncovered colonies of a single species of ant that stretch for 6,000 kilometers (3,700 miles), and a single nest of leaf-cutter ants can contain 8 million individuals. In some tropical forests the biomass of ants is approximately four times greater than that of mammals and, on the global scale, it approximates that of all human beings combined.

But these types of dry statistics had never impressed me terribly, and it was only when I started looking closely at individual ants that I noticed how strikingly beautiful some of them are, and how fascinating their behavior is. The first group of ants that really caught my attention was the leaf-cutter ants (*Atta cephalotes*), which I encountered for the first time in an urban park in San Jose, Costa Rica. On my first morning there I noticed what appeared to be an endless stream of tiny green sailboats floating swiftly over a large buttress root of a tree. On a closer inspection these turned out to be neatly cut pieces of leaves, each carried vertically by a single ant. On some, smaller ants were riding on top of the leaves. The stream of ants continued for about twenty meters, only to disappear underground at the base of a large mound. The first Europeans to see leaf-cutter ants assumed that the leaves were used by ants to build thatched roofs to protect the colony from tropical downpours, and it was not until 1874 that the famous naturalist Thomas Belt realized the true use for the leaves. Leaf-cutter ants are expert gardeners, and the leaves are used as a substrate for growing nutritious fungi (Leucocoprini), the main source of food and water of the ants and their larvae. Until recently this mutualistic symbiosis was thought to involve just these two organisms, but two additional players are now known to be a part of this relationship. Another fungus, a parasitic *Escovopsis*, often invades and can attack and destroy the ants' garden. Usually the parasite occurs at low frequencies, but when the health of the garden is compromised, the parasite becomes virulent, overgrowing the entire fungal crop. Fortunately, ants have a powerful weapon against it, an antibiotic-producing bacteria of the genus *Streptomyces*, which they carry on their bodies and use to control the growth of the parasitic fungus. It is not a coincidence that over half of the antibiotic drugs used by humans are also derived from the same genus of bacteria.

Defending themselves against parasites and predators is what ants do very well. Many species are equipped with a sting, and most have powerful, sharp

Dorylus sp.

mandibles that they never hesitate to use. One time in West Africa I had to strip almost naked to get rid of a swarm of driver ants *(Dorylus)* that invaded, it seemed, every nook and cranny of my body after I accidentally stepped into their marching column. The mandibles of their large soldiers easily cut through skin, and even smaller workers can cause considerable pain. African driver ants and their South American equivalents, army ants *(Eciton)*, do not build permanent nests but instead spend their lives constantly on the move, only occasionally constructing temporary nests, or bivouacs, of their own bodies. These ants are extremely aggressive, and do not hesitate to attack even such a large potential prey as a snake or a careless entomologist. In Africa, entire villages vacate for a few days at the sight of approaching driver ants. In part, the villagers are afraid, but they also know that upon their return their houses will be clean of cockroaches, rats, and any other un-invited inhabitants that the ants consider a meal.

Most ants are opportunistic in their diet, but some are strict specialists. Some species of harvester ants of the genus *Pheidole* specialize on seeds. Their colonies have two castes of workers, small-headed minors and large-headed majors. The minors collect smaller seeds and maintain the nest while the majors' job is to take large seeds and chew them up with their powerful mandibles. A particularly interesting case of a feeding specialization exists in the genera *Leptanilla*, *Adetomyrma*, and *Amblyopone*, in which the queen of the colony feeds exclusively on the he-molymph (blood) of her own larvae, earning the name "Dracula ants" for these genera. The workers and larvae of Dracula ants are far less insidious, and survive on a diet of millipedes and centipedes.

Although the word "ant" is practically synonymous with "tiny," there are some ants that are actually bigger than the smallest species of vertebrates. In Malaysia I was stunned by the unexpected sight of inch-long ants running up and down a tree. This ant, *Camponotus gigas*, is probably the largest ant species in the world, larger than several species of geckos (*Sphaerodactylus*). Equally impressive are bullet ants (*Paraponera clavata*) of Central and South America, which are not only large, but carry venom that can knock a person unconscious. I have never been stung by one, but after ten years of working in Costa Rica, where these insects are very common, I know that it will not be long before one finally gets me. I wonder if it really deserves the title of the most painful, non-lethal animal sting?

Despite the ubiquity of ants, some of their species are faced with the same threats to their survival as other organisms. Nearly 150 species of ants are in-cluded in the IUCN Red List of threatened species. Most of them are considered vulnerable because of very small population sizes or the geographically re-stricted ranges of their colonies. Unfortunately, some of the worst enemies of ants are other exotic ants that were introduced by humans to new habitats. These ants sometimes drive local species to extinction. Only by a stricter control of cargos containing agricultural and forestry products can we slow down the process of homogenization of the world's biota and the subse-quent extinction of countless species.

A winged male bullet ant
(*Paraponera clavata*) (Costa Rica)

127

An army ant worker displays its giant mandibles. The soldiers will attack any enemy, regardless of its size. Some South American indigenous peoples take advantage of the strong muscles of these ants, which do not let go even if the head of the ant is separated from its body, and use them to close up cuts and minor wounds (Costa Rica).

A motionless soldier of the army ant *(Eciton hamatum)* guards an endless stream of workers *(opposite, below)* leaving the bivouac, a temporary nest of these nomadic insects composed entirely of thousands of living ants *(left)*. The bivouac shelters the queen and larvae of the colony and may stay in one place for several weeks. Army ants are relentless predators, capable of overpowering and killing prey hundreds of times bigger than themselves. But a number of insects and arachnids have evolved ways of tricking army ants into accepting them into their colonies. Rove beetles mimic the chemical and tactile language of ants and can travel with raiding ants, gaining protection from predators and exploiting ants for food.

A rove beetle (Staphylinidae: Aleocharinae) is tolerated by army ants by mimicking their chemical signals.

Soldiers of Costa Rican army ants, like all workers in the colony, are nonreproductive females.

When it is time to move the nest in a hurry, older and more experienced workers will often carry young and inexperienced ones to speed up the process.

The arboreal nest of African weaver ants *(Oecophylla longinoda)* is constructed of leaves tied together with strands of silk. The process begins with a number of workers pulling together leaves and holding them in place until other individuals bring the living tubes of glue, their own larvae *(opposite, left and right above)*. Larvae of this ant species have silk-producing glands, and if gently squeezed by workers, they apply it to the surface and edges of leaves, binding them tightly. A nest built of fresh, living leaves provides the colony with a humid, safe environment where they can raise their brood. Often the nest is constructed around aggregations of scale insects, which weaver ants tend and protect, gaining in return sugary honeydew produced by these relatives of aphids. They will also tolerate certain species of caterpillars *(right)* (Lycaenidae and Noctuidae) in exchange for honeydew, allowing them to feed and pupate in the safety of the ants' nest *(opposite, below right)*. But weaver ants are also active predators, hunting animals often many times bigger than themselves. Certain parts of the nest are filled with carcasses of their prey, stored for later consumption. For this reason, weaver ants have been used as biocontrol agents, capable of protecting certain crops from herbivorous insects.

Two green weaver ant workers guard the rear end of a caterpillar (Lycaenidae) (Australia) waiting for a droplet of honeydew to appear from a special gland on the caterpillar's body. Honeydew produced by caterpillars has a high concentration of amino acids, particularly serine.

The arboreal colonies of the Indo-Pacific green weaver ants *(Oecophylla smaragdina)* attract many organisms who profit from the presence of these aggressive insects. Some species of treehoppers (Membracidae) repay by producing honeydew in exchange for protection from predators. A similar mutualistic relationship exists between the ants and caterpillars of some butterflies and moths (Lycanidae and Noctuidae). But a young nymph of the Australasian katydid genus *Polichne* seems to benefit from its resemblance to the ants without giving anything in return. In a phenomenon known as Batesian mimicry, some harmless animals adopt the appearance of noxious or potentially dangerous animals in order to confuse potential predators.

Although generally avoided by insectivorous animals, the green weaver ants are regularly eaten by humans, including Australian aborigines and Dayak tribes in Borneo. Their green abdomens (gasters) have a strong citrus flavor that goes nicely with rice, and their pupae have a creamy taste favored by many in Thailand and the Philippines.

A lone worker guards a treehopper nymph (Australia).

A close inspection reveals that this "ant" is actually a young katydid (*Polichne* sp.).

A nymph of the ant-mimicking katydid (*Eurycorypha* sp.);
close-up of the head below (Guinea)

A worker of the ant *Polyrhachis* sp., a presumed
model for ant-mimicking katydids and spiders (Guinea)

An ant-mimicking jumping spider (*Myrmarachne* sp.) (Guinea)

Can you find the real deal? Only one of the creatures on the page opposite is an ant (hint: ants have elbowed antennae). Many small arthropods take advantage of the general immunity ants enjoy from predators that hunt other insects, and have evolved morphological adaptations to resemble ants in shape and color. Even such unlikely creatures as katydids can be astounding ant mimics as young nymphs, even if they assume more traditional shapes later in life. Jumping spiders also make superb ant mimics, and often use this to their advantage by feeding on their models.

An adult of the ant-mimicking katydid (*Eurycorypha* sp.) looks dramatically different from its nymph (Guinea).

No longer autonomous animals but merely a substrate for growth of a foreign life form, all these insects fell victim to a sneaky pathogen predator—an entomophagous, or insect-eating, fungus. Fungi like these grow inside insects' bodies and dramatically alter their behavior. In many cases the insect is driven to climb tall plants, and shortly before dying firmly attaches itself to the plant with its legs and mandibles. After the host's death the fungus continues to grow within it, eventually producing external fruiting bodies that will disperse new spores.

Bullet ant *(Paraponera clavata)* attacked by *Cordyceps* sp. (Costa Rica)

African ant *(Pachycondyla* sp.) attacked by *Cordyceps* sp. (Guinea)

Weevil (tribe Zygopinae) parasitized by fungus *Stibella buquetii* (Costa Rica)

Unidentified planthopper killed by
Metarhizium sp. (Guinea)

Grasshopper (*Catantops* sp.) killed
by an unidentified fungus (Guinea)

Rainforests are renowned for their spectacular butterflies, but the relatively dimly lit interior of most humid forests makes spotting them rather difficult. In West Africa, the handsome *Euphaedra xypete* is surprisingly difficult to notice when it exposes its metallic blue markings while sitting on a leaf. The same species is equally well camouflaged among fallen, dry leaves on the forest floor, when its folded wings reveal a red and brown pattern, resembling any other half-decayed piece of vegetation. The folded wings of the pale *Salamis parhassus* also make this animal disappear among dead pieces of plants.

In Central America, the owl butterfly *(Caligo eurilochus sulanus)* displays a large fake eye on the underside of its wing. It has been assumed that such markings, common among many species of butterflies and moths, have evolved to mimic eyes of large vertebrates, such as owls or lizards. But it is also possible that their contribution to the survival of butterflies carrying these patterns is that they resemble fungal damage on a dead leaf, making the insect appear unappealing rather than frightening to a potential predator.

Owl butterfly *(Caligo eurilochus sulanus)* (Costa Rica)

Salamis butterfly *(Salamis parhassus)* (Guinea)   Fig eater *(Euphaedra xypete)* (Guinea)

Males of the yellow *Eurema hecabe* gather on a pile of dung they found in a forest clearing (Guinea).

Visitors to humid, tropical locations are often surprised by the number of butterflies and moths aggregating on muddy banks of rivers, mud puddles, and animal dung. Many butterflies and moths are also attracted to human skin and suck sweat or blood from cuts with their proboscis. This behavior, known as puddling, is decidedly more common among male than female butterflies. A series of studies has demonstrated that the main element these insects are after is sodium. Sodium is difficult to find in plant material consumed by the butterfly larvae, although potassium occurs there in abundance. By sucking fluids rich in sodium, adult butterflies try to replace the surplus of the latter element with the former. In extreme cases a moth may imbibe an amount of fluid 600 times its own weight in a single puddling session, expelling the excess water as it drinks and retaining only the precious mineral. It has been shown that a higher sodium content in the male's body enhances his mating success. In addition to sodium, moths and butterflies may extract amino acids from soil, and males of some species contribute these compounds to their mates through a spermatophore, thus investing in their offspring.

In a similar type of behavior, males of the Neotropical clear-winged butterflies (Ithomiinae) can often be seen gathering on dead or withering plants of certain noxious species. These insects extract toxic plant metabolites known as pyrrolizidine alkaloids. The butterflies store these chemicals in their bodies—in part to become unpalatable to predators, but also to use them as precursors to male pheromones needed to attract their mating partners.

Two males of *Pteronymia carlia* gather on a dead plant to collect secondary compounds used both to produce sex pheromones and to make their bodies unpalatable (Brazil).

The West African butterfly *Acraea althoffi* flutters as it drinks seeping water rich in sodium on the edge of a forest river (Guinea).

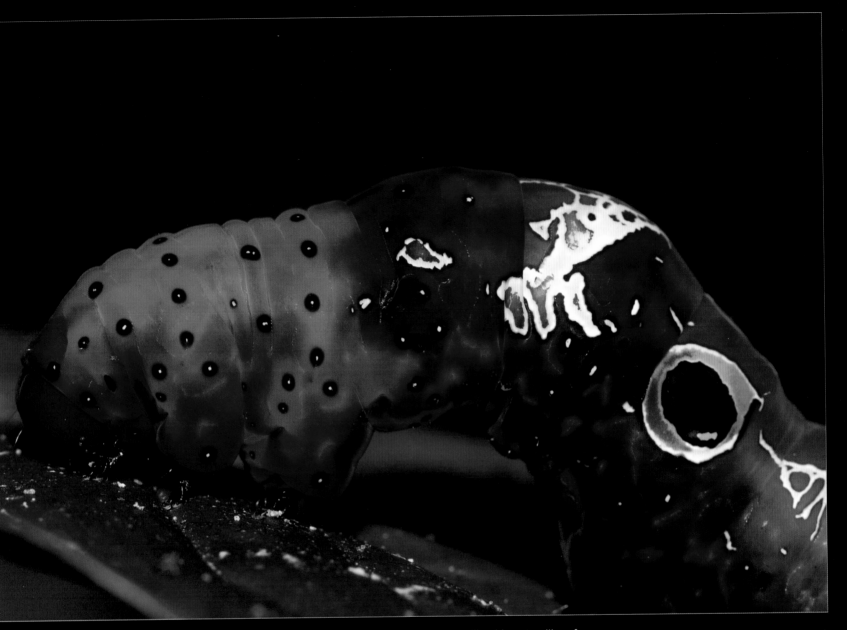

Many caterpillars exhibit false eyespots on their bodies. In some, like this caterpillar of a Costa Rican prominent moth (Notodontidae), the spots are located in the anterior part of the body. The black "pupil" on the red background resembles an eye of some toxic tree frogs, and when found resting on a leaf during the day, this caterpillar may be confused with an amphibian. On the other hand, the white, irregular markings on its body produce a remarkably good imitation of bird droppings splashed on a leaf, and perhaps the false eye looks to a predator more like a seed often found in bird droppings, rather than the head of a frog.

A chrysalis of orion *(Historis orion)* (Costa Rica)

In a phenomenon known as complete development, the majority of insects currently living on Earth go through two distinct, very different stages before reaching adulthood. The first is the larval stage, in which the insect looks very different from the adult and often lives in an environment where the adult would not be able to survive (e.g., water). Most larvae also feed on different types of food than the adult insects. An example of such a larva is the caterpillar of a butterfly. The second stage, the pupa, is a usually immobile phase in an insect's life, during which all its adult organs develop. A butterfly pupa, also known as a chrysalis, looks like a beautiful sculpture, showing all major elements of the adult butterfly's body. The chrysalis is attached to a leaf or branch by a small silk pad, spun by the caterpillar minutes before its final molt. After a few weeks or months, the adult butterfly emerges through a split on the dorsal surface of the chrysalis. Not all butterflies create a free-hanging chrysalis, however. Some butterflies and the majority of moths protect the pupa in a silky cocoon. The cocoon is spun by the caterpillar before its transformation out of a single, extremely long strand of silk. Several species of moths are commercially bred for their silk.

Unidentified Yponomentidae (Costa Rica)

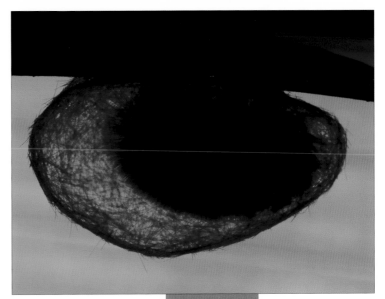

Caligo sp. (Costa Rica)

Unidentified pupa (Costa Rica)

The bright colors of this tiger moth *(Dysschema leda)* advertise its distastefulness to potential predators (Costa Rica).

Rare among moths, this leafroller *(Pseudatteria leopardina)* is diurnal, its bright, conspicuous coloration probably advertising the presence of repellant substances in its body (Costa Rica).

Resembling a military stealth fighter, the body of this hawk moth (*Enyo* sp.) hides powerful muscles that allow it to hover like a helicopter over the flowers on which it feeds (Costa Rica).

Silk moth *(Copaxa syntheratoides)* (Costa Rica)

A cryptically
colored tiger moth
(*Halysidota* sp.)
rests on the
bark of a tree
(Costa Rica).

Night lizard *(Lepidophyma flavimaculatum)* (Costa Rica)

A young Nile monitor *(Varanus niloticus)*

Lizards

A FRIEND OF MINE once sat at the edge of a small lake in Tanzania, admiring a pair of svelte Nile monitor lizards taking their morning bath. A young student at the time, she was pondering her future. For some time she had been thinking of developing a project that would focus on these spectacular lizards. Then a sudden realization came over her—she did not really want to study them, she wanted to be one!

I was struck by her story because I had a very similar experience many years ago in Malaysia, where I witnessed the morning swimming ritual of a different species of monitor. These reptiles can cast a spell on the onlooker, their bodies the very essence of natural beauty. Arguably, few animals can rival lizards' grace. Their bodies, perfected by nearly 150 million years of evolution, exemplify the fact that a good design, once chanced upon, will remain virtually unchanged for a very long time. Most lizards have a similar basic form, with scale-covered skin, four legs, and a long, slender tail. Legs have been lost several times independently in lizards, and one such lineage has become surprisingly successful—we know them now as snakes. The tail has also been modified in a few groups of lizards, becoming prehensile in some, or turning into a flat, leaf-like appendage in others.

Lizards, like all reptiles, are ectothermic—they cannot generate their own heat, and their body temperature is regulated by exposure to sunlight. Not surprisingly, lizards reach the greatest diversity in warmer regions of the world. Tropical rainforests, with their multilayered, complex structure, are home to a stunning array of forms, each perfectly adapted to a particular, unique niche.

The pinnacle of lizards' adaptation to life among trees is, in my opinion, the Southeast Asian genus *Draco*. While visiting a small island off the coast of Thailand, I noticed one day what appeared to be large black and orange moths or butterflies, which every now and then flew from tree to tree. Each disappeared upon reaching a tree trunk, but sometimes a tiny orange flag would suddenly pop up at the site of its landing. It did not take me long to realize that I was looking at some of the world's most amazing lizards, the flying dragons. Although technically their flight is merely a glide through the air, they can cover approximately five meters of horizontal distance for each meter descended. This is remarkably efficient for nonpowered flight, made even more so by the lizard's ability to steer with its tail. The glide is possible thanks to the presence of pataglia—large, wing-like flaps of skin fully supported by the lizard's extraordinarily long ribs. The first two ribs have strong muscles that allow it to control the pataglia. The remaining ribs are connected by ligaments, and they open and close like a Spanish fan. When not in flight, the pataglia fit snugly alongside the cryptically colored lizard's body, making it disappear on the lichen-covered bark of a tree. But males simply cannot stay invisible for too long, and soon after landing they start advertising their presence by flicking a large, triangular dewlap on their throat to attract females and warn off rivals. Oh, how I cursed my broken camera at that moment, lying useless in my backpack. Never again did I have a chance to see these amazing animals.

A few other lizards have also developed the ability to glide from branch to branch, but none rivals the flying dragons. Geckos of the genera *Ptychozoon* and *Thecadactylus* have extensive webbing between their long fingers and large flaps of skin on their tails and bodies, which they use to parachute from trees.

Of course, flying is not the only way to move among branches of the rainforest. In Madagascar and Africa, a steady grip makes chameleons feel secure even on the thinnest of twigs. Unlike primates, whose hands have only one opposable digit (the thumb), a chameleon's foot consists of a pair of opposing pads, one containing two, the other three fingers, enveloped in a mitten of skin. Chameleons' muscular, prehensile tail acts as the fifth limb, capable of holding the weight of the animal as it stretches to reach distant branches below. But in two genera of chameleons, African *Rhampholeon* and Malagasy *Brookesia*, the tail is short and has lost its grasping abilities. These two lineages of lizards have abandoned the arboreal lifestyle and spend their lives among fallen leaves on the forest floor. In an exception to this rule, *Brookesia vadoni* prefers to live high among branches of the bamboos that cover mountain slopes of northern Madagascar. Its short tail struggles to help support the clumsy little animal as it moves rather unsteadily in the vegetation, a reminder that a complex organ, once eliminated by natural selection, cannot be easily regained.

Geckos approach the problem of moving within the vegetation somewhat differently. Instead of trying to grasp the substrate, they cling to it. Special adhesive pads on their fingers (and sometimes also the tail) allow them to move confidently on vertical tree trunks or smooth leaves where other lizards would find little or no purchase. The ability to cling firmly to vegetation and other objects is one of the reasons why some species of geckos occur almost worldwide, inadvertently introduced by humans to many new locations. Another factor that helps geckos colonize new areas is related to their reproduction. Unlike most lizards, some species of geckos do not bury their eggs in the ground but instead glue them firmly to rocks, tree bark, or branches. Combined with the facts that these eggs have hard shells and are remarkably unaffected by seawater, it is easy to see why over time such species, carried by sea currents and rivers, have been able to colonize even the most remote lands.

Sometimes moving to another branch is not an option. A small cat or a snake can easily follow a lizard through the tangled canopy, and in such a situation it is time to jump to the ground. Neotropical basilisks (*Basiliscus*), their Oriental equivalent water dragons (*Physignathus*), and a few other genera of lizards prefer to perch above bodies of water, which both lessens the impact of the fall and also allows them to take advantage of a trick few animals can muster. These lizards have unusually long fingers on their hind legs. Each finger is fringed with small flaps. This increase of the surface area provides enough support to prevent them from sinking if they enter water running at a high speed. Basilisks running across small ponds on their hind legs is a common sight in Latin America, earning them the name Jesus Christ lizards. It seems that the main reason why these and other lizards prefer to run on the surface rather than dive and hide underwater or swim to the other shore when chased by a predator is that fish can also catch and swallow them. By entering the water, these lizards escape terrestrial pursuers, but by moving swiftly above the surface, they evade predators below.

As is the case with all things small, people rarely take time to learn the true facts about lizards, and myths surrounding these animals abound in almost every culture. Although only two species of lizards are venomous (genus *Heloderma*), both occurring only in southwestern deserts of North America, virtually every culture considers at least some local lizard species extremely dangerous or venomous. The more striking the lizard, the worse the prejudice against it. Blue-tailed skinks, a common color pattern among juveniles of many unrelated species, are considered venomous both in the New and the Old World. Some geckos are considered venomous (all geckos are harmless); one South American species is even believed to inject venom with its feet! Chameleons are feared in both Africa and Madagascar, and Mexican Mayas try not to even look at brightly colored *Ameiva* lizards, because it is thought that just a single glance causes severe headaches.

In addition to being singled out for their perceived harmfulness, lizards are threatened by the widespread use of pesticides and synthetic fertilizers in agriculture. Large amounts of these agro-chemicals permeate the environment, and it has been suggested that their presence in the soil interferes with hormone production in the developing eggs of lizards. Pesticides also reduce the natural populations of insects, the primary food source of most lizards. Unfortunately, little is known about the population status of most lizard species worldwide. In Europe many species are already protected, but much more needs to be known about the species inhabiting tropical areas.

Ameiva festiva (Costa Rica)

153

Sharp eyes, lots of patience, and a great deal of luck may reward a persistent naturalist with a glimpse of a helmeted iguana *(Corytophanes cristatus)*. This gorgeous Central American lizard is an example of an extreme sit-and-wait predator, one that spends days sitting absolutely motionless on a tree trunk, waiting for a careless katydid or beetle to pass by. So motionless is this animal that lichens and bryophytes sometimes grow directly on its head, enhancing its already extraordinary crypsis. But like many predators who employ the sit-and-wait strategy, these lizards are capable of an instant switch to attack mode, pouncing swiftly on an unsuspecting victim. They can also run at great speeds, and like their close cousins, the basilisks, they are bipedal, using only their hind legs for locomotion on the ground.

If cornered by a snake or a person, a helmeted iguana inflates its throat and lunges at the aggressor, its mouth agape. In most cases this is a bluff, but nonetheless they are capable of delivering a powerful bite if really irritated. This behavior has earned them a bad reputation among Costa Rican *campesinos*, who fear these harmless lizards, believing them vicious and extremely venomous. Mexican Mayas have a more no-nonsense approach to helmeted iguanas and consider them quite a delicacy.

Hidden in the few remaining patches of Madagascar's natural forest live some of the most extraordinary lizards on Earth. No bigger than a human thumb, dwarf chameleons (Brookesia) are the smallest members of the chameleon family. Unlike their larger cousins, most species of dwarf chameleons rarely climb vegetation, preferring instead to rely on their excellent camouflage while prowling among fallen leaves on the forest floor. Not having to maneuver among thin branches of bushes and trees, their tails have lost most prehensile ability, unlike the tails of other chameleons. Their defensive behavior is on par with their tiny size—whereas larger species of chameleons hiss and bite if cornered, dwarf chameleons adopted the more insect-like behavior of simply folding their legs and dropping to the ground, where they disappear among twigs and debris. If this fails, they vibrate their entire bodies and produce a soft, buzzing noise, perhaps in an effort to mimic stinging insects. Their minuscule dimensions also limit their diet to small, soft-bodied insects, such as termites and springtails.

Dwarf chameleon (Brookesia superciliaris)

Dwarf chameleon (*Brookesia vadoni*)

Geckos do not have typical eyelids and cannot close their eyes. Instead, its eyelid forms a transparent shield, which needs to be cleaned regularly with the gecko's large, muscular tongue (*Uroplatus phantasticus*, Madagascar).

Finding a leaf-tailed gecko *(Uroplatus)* one night in Madagascar as it slowly moved from branch to branch using its hand-like feet was one of those reaffirming moments that reminded me once again why I chose to become a biologist. I gently picked it up from the tree; it did not try to escape, perhaps confused by the light of my headlamp. It felt warm and soft in my hand, as if enveloped in velvet. The skin of geckos is covered with tiny, bead-like scales that give them a soft, delicate feel. Their skin is also their main line of defense—if caught by a snake or bird, geckos easily shed large patches of skin, with the lizard slipping away relatively unharmed. The exposed areas do not bleed, and they heal very quickly, regrowing new skin and scales. Like many other lizards, a gecko can also reject its tail to confuse and escape from a predator.

The long, wide fingers of the gecko held firmly onto my hand. They felt slightly sticky, yet dry. Their unusual fingers give geckos the ability to walk on smooth, vertical surfaces, including glass. Many explanations had been proposed for this phenomenon, but only recently it was discovered that the mechanism responsible for it lies at a molecular level. Each digit of the gecko's foot has an adhesive pad composed of millions of hair-like setae. The tip of each seta ends with tiny terminal branches, or spatulae. Thus, a gecko's foot has about a billion microscopic points that bind to individual molecules of the substrate using what is known as van der Waals intermolecular attractive forces, allowing them to walk, even upside down, on smooth branches and leaves.

Leaf-tailed *Uroplatus* geckos occur only in Madagascar. Many of them cling to the last remaining patches of the natural forest of this island, and although no species of this genus is officially listed as threatened or endangered, there is little doubt that they will survive only as long as their quickly vanishing habitat. I put my gecko back on the branch and it soon resumed its deliberate search for small frogs and insects.

Fantastic leaf-tailed gecko
(*Uroplatus phantasticus*)

West African gecko (*Hemidactylus fasciatus*) (Guinea)

Leaf-tailed gecko *(Uroplatus sikorae)* (Madagascar)

Geckos are the most common lizards to be seen at night in a tropical forest. They are also the only lizards to be heard at night. Many species vocalize loudly, some even chorusing like frogs or katydids. All geckos have excellent night vision, allowing them to stalk nocturnal insects and small vertebrates. Its stalking behavior is remarkably similar to that of cats—not only does it slowly sneak up on its prey before striking, but it also waves its tail like a cat hunting a mouse.

During the day most geckos rest, blending perfectly into their surroundings thanks to their cryptic coloration. Their body color can be changed thanks to the presence of chromatophores, special cells that by contracting and expanding can alter the color and pattern of the skin. Many geckos like this *Uroplatus sikorae* from Madagascar have their body covered with small flaps and fringes that further distort the familiar outline of a lizard and meld it with the surroundings.

Hispaniolan giant gecko *(Aristelliger lar)* (Dominican Republic)

Fantastic leaf-tailed geckos can change their color easily thanks to the presence of chromatophores in their skin (Madagascar).

Mouth agape, the giant leaf-tailed gecko *(Uroplatus fimbriatus)* tries, unsuccessfully, to stop me from taking its picture. The enormous mouth of this species is fitted with more teeth than that of any other land animal, perhaps to help it capture frogs and other small, slippery animals that constitute its diet (Madagascar).

The fantastic leaf-tailed gecko *(Uroplatus phantasticus)* becomes virtually invisible during the day, thanks to its leaf-like shape, coloration, and posture, assumed to break the symmetry of its body.

These flies in the rainforest of Costa Rica are all males, engaged in a communal courtship display known as leking.

Another example of leking behavior of males, this time of the stalk-eyed flies (Diopsidae) in the rainforest of Guinea

Tse-tse fly (*Glossina* sp.) (Guinea)

An unidentified fly (Muscidae) (Guinea)

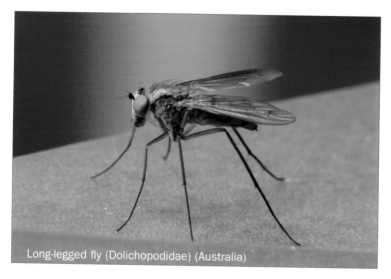

Long-legged fly (Dolichopodidae) (Australia)

Stilt fly (Micropezidae) (Guinea)

Flies do not top the list of most people's favorite animals, and I must admit that I often find myself infuriated with overly friendly tse-tse flies or mosquitoes. Yet, like it or not, flies are exceptionally important members of tropical ecosystems, fulfilling functions that include waste removal, soil production, water filtration, plant pollination, and last but not least, being food for other organisms. This large and very diverse group of insects includes many parasitic species.

Stalk-eyed fly (Diopsidae) (Guinea)

My first reaction to the sight of a mosquito feeding on the blood of a sleeping anole was to flick it off its head. But are parasitic flies really that bad? They may be unpleasant or even harmful to an individual host organism, but mosquitoes and other parasites play an important role for the ecosystem as a whole, regulating the hosts' population sizes and preventing them from explosive growth. Flies that went along the path of obligatory parasitism show some remarkable adaptations to the parasitic lifestyle. The most common is the loss of wings, exemplified here by *Forcipomyia* sp. (Ceratopogonidae) *(opposite, below)*, permanently attached to the antenna of a walking stick (*Metriophasma* sp.). Fly families associated with mammals are often heavily sclerotized and flattened, preventing dexterous mammals from easily picking the flies off their fur. Members of the family Nycteribiidae are often found on fur and wing membranes of tropical bats *(opposite, above)*. As for the mosquito on the anole's head, I decided that it was unfair to swat mosquitoes on myself but let them drink the blood of my photographic model, so I brushed it off without waking up the lizard.

A parasitic fly (Nycteribiidae) on a wing membrane of a fruit bat (Guinea)

A flightless, parasitic fly (*Forcipomyia* sp.) attached to the antenna of a walking stick (*Metriophasma* sp.) (Costa Rica)

Mantids

Sex and death—a combination of these two elements seems to sum up our image of preying mantids. For thousands of years these insects have achieved an almost cult status in many cultures, including the Western ones. (Actually, the cult was real in ancient Egypt, and preying mantids were considered minor deities.) Our fascination centers on the mantids' mating behavior, during which a cannibalistic female devours her ill-fated partner. Add to this the almost human-like body form of the mantid, with a long "neck," a movable head with big eyes, and piously folded front legs, and it is not at all surprising that these ostensibly singular insects have enjoyed an inordinate amount of attention. Therefore, to many it may be quite disheartening to learn that the mantids' notorious cannibalism is actually a rare phenomenon rather than the rule, observed occasionally only in a handful of species. Furthermore, mantids, these charismatic predators, appear to be simply a lineage of highly evolved cockroaches. But regardless of the true nature of their sex lives or less-than-popular relatives, preying mantids are truly fascinating animals. To me they epitomize the splendor of the tropics and the magnificent evolutionary forces that created it. Growing up in Poland, a country where within the last few hundred years preying mantids have gone extinct nearly everywhere, I longed for places where free speech was allowed and mantids were plentiful.

There are about 2,300 described species of preying mantids, most of them concentrated in the circumtropical belt of the globe. A few species from Europe and Asia have recently become nearly cosmopolitan, introduced to new areas with seedlings or shipments of fruit. For a group as popular with the public as mantids are, we know shockingly little about the biology and behavior of nearly all of their species. Most entomology textbooks rehash anecdotal information based on observations of a handful of species common in Europe or North America. Tropical forms are mentioned only in passing, if at all. Fortunately, the last fifteen years have seen a renewed interest in this taxonomic group among scientists, bringing to light new and quite unexpected facts about mantids.

Preying mantids, as the name implies, prey on other organisms (an alternative spelling, "praying mantids," is sometimes used, emphasizing the "begging" posture of these insects). They employ three types of hunting strategies. Generalists such as arboreal *Liturgusa* prefer hunting on smooth, unobstructed tree trunks. They camouflage themselves among lichens, but will burst into a high-speed pursuit of any prey that passes their field of vision. Cursorial hunters are common in open, grassy habitats. African *Hoplocorypha* is an example of species that patrol large areas of dry terrain on their long, slender legs, looking for grasshoppers and other insects. But most mantids belong to a group known as ambush hunters. These species are sit-and-wait predators, rarely compelled to move from their perch, relying instead on a chance encounter with a suitable meal. Some species, like the southern African *Pseudocreobotra wahlbergi*, mimic buds and flowers and preferentially wait on plants visited by pollinators. Other species also tend to wait in "high traffic" areas where insects move to and fro. No species of mantid is known to seek out any particular kind of prey, and only the size and strength of their raptorial front legs limit the size of their meal. Large mantids have been seen feeding on lizards, hummingbirds, a huge variety of insects and other arthropods, and, of course, other mantids. Cannibalism is not restricted to reproductive circumstances; the first meal of a newly hatched mantis is often its own sibling. This indiscriminate voraciousness is the reason why mantids never occur in large groups, and attempts to use them as natural pest control agents invariably fail.

Interestingly, it has been demonstrated recently that some species of mantids supplement their diet with pollen, a nutritious source of protein. Young nymphs of *Tenodera aridifolia* can subsist exclusively

on pollen, and individuals who were able to supplement their insect diet with it showed significantly higher survival rates than their siblings restricted to eating just insects. Adult mantids have never been observed to feed on pollen, but they often eat bees, including their pollen sacks.

Although reproductive cannibalism is not as common as it was previously believed, a successful courtship is still a tricky affair for a solitary, silent insect living in dense tangles of rainforest vegetation. Anecdotal observations suggest that mantids may employ chemical signaling in their courtship, but until very recently no solid proof of pheromonal communication in these insects existed. Only recently a pheromone consisting of a mixture of compounds known as pentadecanal and tetradecanal has been isolated from a West African *Sphodromantis*. A series of ingenious experiments has unequivocally shown that this pheromone acts as a powerful sexual attractant.

But sensing the scent of a receptive female and getting to her are two very different things. Mantids are generally large, slow-flying diurnal insects prized by lizards, birds, and other visual predators. Thus, many diurnal species prefer to mate at night, when fewer eyes can notice a large insect flying slowly toward his mate. Unfortunately, there are some predators that rely on hearing rather than vision in their quest for an insect meal. Bats use echolocation, which involves bouncing short, high-frequency clicks off objects in their flight path, creating a detailed map that includes even the tiniest insects unfortunate enough to cross that path. Luckily, millions of years of coexistence have put enough selective pressure on flying mantids to result in the evolution of a unique hearing organ that helps them detect bats' high-frequency signals. They can now perform a life-saving flight maneuver. Between the coxae ("hips") of the mantid's last pair of legs lies a deep groove, at the bottom of which a pair of closely spaced membranes forms the basis of the insect's single ear. While technically a paired structure, the hearing membranes are spaced so closely together that directional hearing is not possible. Thus, when in flight, mantids cannot determine where a bat's ultrasonic clicks are coming from, hence they employ a strategy similar to that seen in moths and other insects: they perform an evasive maneuver in a random direction and quickly drop to the ground. This strategy is surprisingly effective. In addition, it has been shown that mantids' hearing is finely tuned to the specific frequencies produced by bats most common in their area.

No species of preying mantids has any significant economic value, and perhaps this is one of the reasons why research on this captivating group of insects has lagged behind other insect orders. The recent revival of interest in mantids is a welcome sign that will certainly lead to more interesting discoveries. We also need to look closely at threats faced by these animals from various human activities—habitat loss always being the most important. Currently only one species of mantids is listed by the IUCN on its Red List of threatened species, but there is no doubt in my mind that many more may be vulnerable or endangered. New information on mantids will help us identify species in the greatest need of protection before the situation becomes critical.

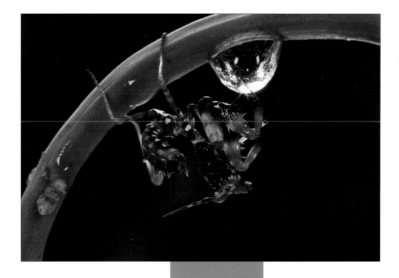

Sit and wait, and something will come. This is the strategy employed by most preying mantids in the lowland Atlantic forest of Costa Rica. The female of giant mantid *Macromantis hyalina* devours a sylvan katydid *(Idiarthron hamuliferum),* a feast that will keep her satiated for only a short while. She needs a lot of protein to produce her large case of eggs.

Some long-lived species of mantids were recently discovered to have live mosses and bryophytes growing on their bodies. In one individual, several species of these tiny plants were found on its body. A similar phenomenon was also observed in a species of walking stick.

A small, unidentified mantid waits for an unlucky insect to land near her on a clump of ferns (Guinea).

A male of the large mantid *Tauromantis championi* gingerly positions himself as far from the female's grasping legs as possible. Although his chances of surviving this mating are high, he does not take any unnecessary risks. It has been shown that males will preferentially mate with recently fed females, perhaps by being able to detect the scent of her last meal. Cannibalism during mating is rare, and often induced by a disruption, such as the arrival of another male. But even when decapitated the male continues to transfer his sperm to the female's genital opening, his sacrifice not entirely wasted.

Mantids are not fussy when it comes to prey. They will try to catch and devour any animal small or weak enough to be held in their front raptorial legs. The elements of the leg that constitute the grasping mechanisms are the femur and tibia, whereas the foot plays no role in hunting, and is used only when climbing vegetation.

Hooded mantid *(Choeradodis rhombicollis)* (Costa Rica)

An unidentified mantid
(Dominican Republic)

African mantid (*Sphodromantis* sp.)
(Guinea)

Vertical lichen- and moss-covered tree trunks are the hunting grounds of a number of preying mantids. Their bodies are generally flat and cryptically colored. Tree mantids are unbelievably fast, running up and down the trunk at blazing speeds chasing crickets, moths, or any other small animals they can spot. But they are not the only hunters on the tree, and this Costa Rican *Liturgusa annulipes* must compete against equally well-camouflaged and fast Selenopid crab spiders (Selenopidae).

In Guinea, the mantid *Theopompella heterochroa* flattens itself against the bark. Irregular patches on its wide wings help it mimic lichen.

Dirt-free feet are critical for running on smooth, vertical surfaces, and mantids clean them frequently *(Theopompella heterochroa)* (Guinea).

177

It may seem impossible looking at the almost human-like features of a preying mantid, but these insects are most closely related to cockroaches. They share with them all the basic features: the triangular head, large pronotum, legs with strongly elongated coxae (hips), and many others. And yet no two groups of insects could differ more in their behavior. Diurnal, predaceous, and mostly sedentary, mantids have lost many characteristics we usually associate with cockroaches, such as the love of darkness and agility.

Large species of mantids spend days hanging upside down from branches, waiting for a suitable prey to saunter by. This position relieves their leg muscles from the need to support their heavy bodies, saving energy. Many mantids will only face up when moving to a higher ground or when mating.

SAVANNAS

I HAD NEVER REALLY BELIEVED in being able to "feel" somebody's stare on my back until I found myself standing in the middle of a southern African savanna, surrounded by a sea of grass that stretched up to the horizon in all directions. I have no problem with being alone at night in the rainforest, but African savannas, be it day or night, make me feel really vulnerable and exposed. It is not just the lions and hyenas that my imagination puts behind every bush and termite mound. It is also the inability to see what lies near my feet, potentially a thick, muscular coil of a puff adder or a spitting cobra. I had been put in a hospital by a viper in the past, and knew that a snakebite in the middle of Botswana could be the last thing that ever happened to me. The technological highlight of Rakops, the nearest town, was a crank-operated gas pump, and even that was separated from our camp by a two-hour drive on bad, sandy roads. I doubted that Rakops had a medical facility capable of administering antivenin and treating anaphylactic shock. Why then did I go there, dragging along my unsuspecting wife? The reason for my fascination with tropical savannas is the fantastic diversity of animal life in these ecosystems, much of which shows astonishing adaptations to the seasonal fires that regularly happen there.

Savannas receive less rain than tropical forests at the same latitudes, ranging from 500 to 1,500 millimeters per year. They are seasonal, with a pronounced dry season that in extreme cases can last for more than half of the year. It is often at the end of the dry season that savannas fall victim to sweeping fires. Fire is not a force usually associated with life, and yet many organisms in grasslands depend on it for their reproduction, while others have evolved ways of surviving and benefiting from it. It usually starts with a strike of lightning at the end of the dry season. As the flames turn stalks of grass into strands of wispy ash and scorch the bark of trees scattered on the plain, thick seed husks of fire-dependent plants crack from the high temperature, freeing the plant embryos to sprout as soon as the ground has cooled. Rejuvenating rains that usually follow will allow the ash covering the ground to release its mineral contents into the soil, providing new plants with nitrogen, phosphorus, and other elements they require for healthy growth. But before the plains become covered with a lush green carpet of new plant life, the once grass-covered expanse is black. A green or yellow grasshopper that miraculously survives the fire suddenly finds itself exposed, and soon falls prey to a bird. Yet only a couple

of weeks later the scorched ground is alive with insects munching on sparsely scattered, juicy new blades of grass. This time, however, the grasshoppers blend perfectly into the background, thanks to a phenomenon known as fire melanism. Although the exact mechanism that causes some insects to turn black following a fire is not completely understood, physiological changes induced by differences in the humidity and perhaps also the chemical composition of the substrate cause many insects that normally display light coloration to darken following a molt. Other insects such as moths cannot change their coloration during their life, but have evolved a strategy that allows at least part of the population to survive a period when their normally cryptic coloration suddenly makes them easy to pick out. In such species, populations often display a strong color polymorphism. Some individuals have coloration that makes them disappear against the background of grass or light tree bark, but others are black or darkly colored. These dark individuals inadvertently act as a safety reserve for the species, allowing it to survive periods when many of its members are at a heightened risk of predation. This is not to imply that color polymorphism evolved for the explicit purpose of helping the species to survive—the evolutionary process has no predetermined goal. Any effects of natural selection we see in living organisms are by-products of the competition between alternate copies, or alleles, of genes found in the organisms' populations. The allele that enhances the survival of its bearer has a greater chance of being passed on to the next generation, increasing its frequency among the members of the species.

Fire is a major force that shapes the life of savanna ecosystems, stripping the soil of nonwoody vegetation, but for most of the year grasses dominate the landscape. In open savannas the grass cover is complete or very sparsely interspersed with other vegetation. But up to 75 percent of a savanna can be covered by trees and still be classified as such, although the term "parkland" is often used to describe grassy areas with a large percentage of tree coverage, especially if they occur at mid-elevations. Savanna trees such as the common African mopane (*Colophospermum mopane*) often exhibit strong fire resistance. Their bark is thick and corky, and the cells of these plants contain crystals of calcium oxalate. When the temperature of the wood during a fire reaches 370°C, this substance starts to decompose, producing copious amounts of carbon dioxide that acts as a powerful fire retardant. In addition to reducing fire damage to the tree, calcium oxalate is a natural repellent against termites and other wood-boring insects. Ironically, the fire resistance of some savanna trees is also the reason for their rapid disappearance from this habitat. Because they burn much slower than other species, such trees are sought as an ideal wood for cooking, providing hot and long-lasting fire.

Savannas, despite their relative dryness, have exceptionally high primary productivity, rivaling or even exceeding that of forest ecosystems. As a result they also show surprisingly high animal species diversity, despite being distinctly less botanically diverse than forests. Grasses, which dominate savanna ecosystems, contain much lower levels of secondary compounds than tropical forest tree leaves, making them more palatable to a wide range of herbivores. In addition, dead plant material is very quickly recycled by termites, which are unique among insects in their ability to digest cellulose, owing to the presence of symbiotic protozoans in their digestive systems. The termites' activity allows for the nutrients locked in dried grass stalks and fallen leaves to be quickly returned to the soil, resulting in their almost immediate availability to the next generation of plants. Thus, savannas always have large amounts of plant material available to grazers and other herbivores, and large numbers of species feeding on the same plants can coexist.

Insects of many different species that live in the savanna grasses have evolved to blend well among the shapes and colors of those grasses. As a result, diverse species resemble each other as well. Many grasshoppers, mantids, and other animals have evolved long bodies that blend well among stalks of grass. Some species have developed disruptive coloration, which breaks up the insect's outline into less recognizable shapes. For example, the grasshopper of the genus *Leptacris* is marked with a silver reflective strip along the sides of its long, cylindrical body. This pattern not only divides its already slim contour into two unequal halves, but also shows a certain degree of reflectance, assuming the hue of the insect's immediate surroundings. Many insects turn their bodies into perfect replicas of blades of grass through the presence of long, conical structures on either their heads or the tip of their abdomens. In addition to the morphological resemblance to plants, a significant number of grass inhabitants have converged on a similar way to move, adopting a shaky, seemingly hesitant gait, a perfect imitation of a narrow piece of vegetation trembling in a slight breeze.

Like most natural tropical ecosystems, savannas suffer from uncontrolled human activity. Grazing by domestic animals reduces the grass coverage, which lowers the frequency and extent of natural fires. This in turn allows species of plants that normally do not occur in the savanna habitat to colonize some of its area, a phenomenon known as bush encroachment. This, combined with the introduction of alien plant species, leads to a significant lowering of the overall biodiversity of the savannas. Unchecked harvesting of firewood and clearing of savannas for cash crops such as sugar cane also cause a steady decline in the areas covered by this habitat. On the other hand, human activity may lead to the appearance of savannas in parts of the world where historically they have never been. With the exception of small areas of Nicaragua, Honduras, and Belize, where natural savannas can be found, savannas in other parts of Central America are the result of cattle ranching and regular burning of the forests by people. This has led to the establishment of large, open areas that now support an extraordinary diversity of birds and other animals, creating a potential for their use in ecotourism.

The shrubby
vegetation of
southern African
savannas is full of
interesting small
animals. Many, like
the acacia katydid
(Terpnistria zebrata),
exhibit a characteristic
color pattern, mixing
green and white ele-
ments, which make
them disappear
among highlights
and shadows of
small-leafed bushes
illuminated by the
bright sun.

Armored ground katydid *(Acanthoplus discoidalis)* (Botswana)

Flap-neck chameleon *(Chamaeleo dilepis)* (South Africa)

Mole cricket *(Gryllotalpa africana)* (Botswana)

Bell cricket *(Homoeogryllus orientalis)* (South Africa)

An unidentified cricket killed by an assassin bug (Australia)

A female tree cricket (*Oecanthus* sp.)
deposits her eggs in a plant stem (Guinea).

Tropical savanna nights are often louder than the day. The culprits are many hundreds of species of crickets, each with a unique, pure-tone call, that together create unforgettable soundscapes, different for each region. In their calls, crickets use frequency modulation: the frequency of the sound changes within each individual chirp, creating a song similar to that produced by birds, albeit the chirps are often too fast for the human ear to appreciate the similarity. Crickets' close relatives, katydids and grasshoppers, use time modulation in their calls, creating more monotonous, noisy calls.

Some crickets use sophisticated techniques to amplify their calls and make them carry for surprisingly long distances. Tree crickets *(Oecanthus)* chew holes in leaves and, while calling, males position their wings in such a way that the leaf acts as an additional membrane to amplify the call. Male mole crickets *(Gryllotalpa)* call from underground horn-shaped burrows that open skyward, directing the sound toward flying females. But great efficiency in sound production is a double-edged sword: although male crickets can attract females from many miles away, they also attract conspecific males, who wait near the singer and intercept the female before she reaches the legitimate suitor. A number of parasitic flies, other insects, and lizards are also attracted to crickets' calls for less-than-romantic purposes.

Savanna katydid (*Anoedopoda* cf. *lamellata*), possibly a new species from Guinea

Senegal katydid *(Thyridorhoptrum senegalense)*

Afro-montane savannas of West Africa resonate with loud, vibrant songs of scores of grass-feeding katydids. The songs of males serve a dual purpose—they try to attract females from as far as a mile away and at the same time they advertise to other males that their territory is taken. Each species has a call that is distinct in either its frequency or the pattern of individual chirps. Thus, the acoustic space of the savanna is precisely divided, with little or no overlap of the sound signals. It has been demonstrated that if two species of katydids that produce very similar calls and thus signals likely to interfere with each other occur in the same habitat, one of them will change its calling habits. It will start calling at a different time of day, when its competitor is silent. For example, it may start calling during the day, even if the preferred courtship time of its species is night.

Because a female will never approach a male producing the "wrong" call, the call acts as the first and a very effective barrier against mating with a different species, or hybridization. This strong barrier may result in a relaxation of other, more typical barriers to cross-species mating, such as the morphology of the katydids' reproductive organs. Some genera of katydids such as *Ruspolia* or *Pseudorhynchus* have a number of so-called cryptic species—morphologically indistinguishable from each other but very different in their acoustic behavior.

The song of a katydid is produced by the sound-producing, or stridulatory, organs located at the base of the front wings of the male. On the left wing, which in katydids always overlaps the right one, one of the veins takes on the role of a bow. Its lower surface is equipped with a series of small teeth or pegs, and with each opening or closing of the wings these are drawn across the edge of the right wing, causing it to vibrate. This vibration is further amplified by a large, clear membrane on the left wing, acting much like a speaker in a home stereo system. Sometimes even the right wing has a large membrane, resulting in a very loud call.

Conehead katydid
(Pseudorhynchus robustus)   191

A female pollen katydid eats her partner's spermatophore.

Although Australian nature has been made famous by its marsupial and avian members, its insects have always attracted me to this continent. Australia is brimming with unique lineages in almost any group of organisms, but few have as many biological gems as the katydids. During the last decade pollen katydids *(Kawanaphila)* have fascinated many biologists. These slender animals are covered with fine, soft hair that make them ideal dispersers of the pollen of plants they feed on. Unusual for katydids, their mouthparts are strongly elongated and perfectly adapted to fit into flower corollas. But it is their courtship behavior that makes them especially interesting. Males of pollen katydids produce particularly large and protein-rich spermatophores, and when food resources are scarce, they gamble all by the right or wrong choice of a mating partner. This causes the males of *Kawanaphila* to become coy and choosy during the mating game, and many females compete for access to a single male. Not surprisingly, the female who can hear and locate the male first has the greatest chance of mating with him. This has led to the evolution of large ears in females, much larger than those in the males, and it has been shown that females with the largest ears are also those that enjoy the greatest mating success, reaping the benefits of a large, nutritious spermatophore.

Balloon-winged katydids (Tympanophorinae) show unusually strong sexual dimorphism. Males sport flamboyant green and red coloration and have large, fully developed wings. Females *(left)*, on the other hand, are drab, completely wingless, and somewhat lizard-like in their appearance. But both sexes are equally efficient predators, capable of jumping considerable distances in their pursuit of insect prey.

A balloon-winged
katydid nymph
*(Tympanophora
uvarovi)* (Australia)

Termite colonies have a complex social structure, with each individual fulfilling a specific role. Unlike ants and bees, in which all workers are females, workers in the termite colony belong to both sexes, albeit they also cannot reproduce. A single reproductive female *(upper left)* is responsible for laying all the eggs, and she can produce up to a thousand of them in a single day. According to some circumstantial evidence, a termite queen can live for seventy years or perhaps longer, thus producing an almost unimaginable number of offspring. Reproductive males of termites are also long lived and stay with the queen, mating with her multiple times.

The division of labor within a termite colony is well defined. Young nymphs stay in the nest until they are capable of foraging outside, while workers tend the eggs *(upper right)*, forage for plant material *(lower left)*, and conduct all activities related to the construction and maintenance of the nest. In some species a morphologically distinct caste of soldiers defends the colony with powerful mandibles *(lower right)* or a special gland on the head that can squirt repellent chemicals at an intruder.

Termites are some of the very few insects capable of digesting cellulose, the main component of plants' skeletal structures. They accomplish this by providing a home for symbiotic protozoans, which live in each termite's hindgut.

Termites are a major force shaping the plant composition of many savannas, and their large mounds are a common sight in most open areas of the tropics, such as this savanna in northern Australia. The primary function of these often enormous edifices is thermoregulation and gas exchange for the colony, while breeding, raising young nymphs, gardening of fungi, and most other activities happen in the underground portion of the nest. The rock-hard walls of the mound are constructed from a mixture of soil, saliva, and feces of worker termites. Termite mounds are used by a wide range of animals, such as birds, lizards, and small mammals, who use these protective structures to nest, breed, or estivate during the dry season.

Termites play a critically important role in the functioning of open habitats by recycling dead wood and other plant material. Their underground tunnels improve the porosity and aeration of the soil, creating better conditions for plant growth. Termites also help reclaim soils damaged by overgrazing in many arid regions of the tropics, and their positive role in many threatened habitats of the planet almost makes me want to leave alone the colony munching the basement of my house.

African savannas have an almost fractal organization, with similar types of interactions repeating themselves at different spatial scales. Large herds of ungulates graze and browse the plains, and predators such as hyenas, jackals, leopards, or lions use various hunting strategies to kill and devour these herbivores.

The near vicinity of each termite colony replicates these interactions, albeit in dimensions thousands of times smaller. The termites stand in for the plant-feeding antelopes and gazelles, harvesting seeds, fresh grass, dry stalks, and any other plant material they can find. In those species in which a soldier caste is present, soldiers guard the entrances to the nest and try to ward off predators, like a female antelope protecting her young. In other species the workers are more vulnerable and a whole suite of predators awaits their emergence from the ground. Some, such as assassin bugs *(Lopodytes),* are ambush predators, waiting motionless like a leopard only to jump on a worker unlucky enough to get too close *(above left).* Sand-diver spiders *(Ammoxenus),* much like cheetahs, chase workers at a very high speed, and when successful, drag their prey to a safe spot to devour it *(below left).* The hyena's approach to hunting is adopted by harvester ants *(Pheidole),* which will attack a much larger termite as a group and cut the poor animal into pieces *(opposite, above).*

Grasshoppers

ONE OF THE DIFFERENCES between African or European grasslands and those in the Americas and Australia is that the former are significantly noisier during the day. The culprits are numerous species of slant-faced grasshoppers (Gomphocerinae), one of few groups of these jumping insects that have developed a widespread use of acoustic signaling. Although there are some singing grasshoppers in the Americas and Australia, nowhere have they achieved such a remarkable diversity as in southern Europe and Africa. Unlike their distant relatives, the crickets and katydids, which produce sound by rubbing their wings together, grasshoppers do so by striking their hind legs against protruding veins on their wings or ridges on the abdomen. The calls of grasshoppers are rather soft and buzzing, but large choruses of males can make a meadow resonate with a very pleasant medley of songs, a sound that to me represents the very essence of summer.

But many people have less than pleasant associations with grasshoppers. Some of these insects are responsible for devastating damage to crops, and millions of people are affected every year by damage caused by migrating hordes of locusts. There is no denying it—pest grasshoppers have a serious negative impact on our global economy, and for the foreseeable future their control will be a central issue for pest-control agencies worldwide. But what causes them to have such a destructive effect on agriculture, especially in dry areas of Africa and Australia? The answer lies in a peculiar aspect of their biology, which, under certain conditions, causes some species of otherwise harmless grasshoppers to turn into ravenous eating machines that band together and destroy everything in their path. Most pest insects are products of human activity. We often unintentionally cause species populations to explode because our intervention, such as the creation of monocultural crops or forests, has removed their natural enemies or other factors that limit their population growth. But locusts are a very interesting case of a "natural pest," an organism that causes widespread damage regardless of the presence of man.

It usually starts with a long-awaited rain, which can quickly transform an inhospitable, dry terrain into a green pasture, covered with soft, lush vegetation. It is a feast for the desert locust (*Schistocerca gregaria*), an African grasshopper used to gnawing dried-up, dusty stalks and grasses. Females respond with a production of larger-than-usual eggs, and young nymphs thrive and mature rapidly. As the number of insects in the population grows, so does the chance of physically running into another member of their own species. Sensitive hairs on the outer surface of the insects' femora trigger the endocrine system of young grasshoppers to produce hormones that will drastically change their appearance and behavior. Soon these green, cryptically colored insects turn vividly yellow and black. They get bigger and their muscles grow stronger than usual. They no longer shy from one another but seek each other's company and start forming dense clusters, feeding and marching together. The adults are also different now: from sandy gray they turn bright yellow, their wings become longer, and the body larger and more streamlined. Males start producing pheromones that accelerate the development of other individuals, leading to synchronized maturation across the entire population. At the same time females' pheromones attract other females, causing them to lay eggs close to each other in dense groups. The entire population grows rapidly and soon the transformation into a migratory phase is complete. This transformation from a solitary into a migratory phase with different morphology and behavior is what defines locusts. Many species of grasshoppers occur in large numbers, but only a few undergo this phase transformation.

Once all members of the population have matured, the desert locusts are ready to move on. Shortly after their final molt, the fully winged adults start flying erratically. One group of flying adults stimulates others to take wing, and suddenly an enormous swarm develops and moves off in the direction of the wind. The size of a single swarm can be larger than any other single congregation of organisms on the planet. Desert locust swarms can range in size from 100,000 to 10 billion insects, making the swarm, in terms of the number of individuals, larger than the entire human population. These giant, live clouds can stretch from a mere square kilometer to about a thousand square kilometers and weigh more than 70,000 tons. The amount of food such a mass of insects requires is staggering: in one day a very large swarm can devour the equivalent of food consumed daily by about 20 million people!

Exceeding the carrying capacity of the environment appears to be the principal factor responsible for the swarming behavior of locusts. Massive migrations have evolved in response to diminishing food resources in an attempt to find and colonize new habitats. The strategy turned out to be exceptionally successful, and desert locusts are now widespread over a great area covering most of Africa and a large portion of Southwest Asia. Their ability to stay airborne, carried by winds over huge distances, may also be responsible for the occurrence and subsequent adaptive radiation of the genus *Schistocerca* in the Americas. In Australia the plague locust (*Chortoicetes terminifera*) causes great damage to crops, while tropical Africa, Madagascar, and some areas in Asia fall victim to the migratory locust (*Locusta migratoria*).

But not all grasshoppers are harmful. In reality, most of them have a very positive effect on their habitats, and other species of animals depend on the presence of these insects. Some species of grasshopper previously believed to compete with grazing mammals help them by feeding on species of plants unpalatable to ungulates and promoting growth of plants favored by these mammals. It has been demonstrated recently that the presence of grasshoppers in cattle pastures significantly increases cattle productivity. Huge populations of grasshoppers in African grasslands also allow migratory birds from Europe and northern Asia to survive the winter, and many reptiles and amphibians rely on grasshoppers as their main source of food. Even humans include grasshoppers in their diet. During a collecting trip in Zimbabwe I was at first surprised, then irritated by the fact that all grasshoppers brought to me by my local helpers were missing their jumping legs. Only after watching them expertly catch these insects and instantly pull off their appendages did I realize that they thought I was going to eat these insects, and they were doing me a favor by discarding "useless" parts. Grasshoppers are an important part of the diet in rural and suburban areas of Zimbabwe as well as in other regions of Africa, the Americas, and Asia. I strongly believe that the only reason we do not eat these insects in the United States and Europe is because of the lack of large-bodied species of grasshoppers occurring in large, predictable populations. If they were present, practicality would quickly overcome any cultural biases against eating insects. Grasshoppers make a very tasty and nutritious meal, and with a mere three pairs of legs should be considered less repulsive than shrimp, which have five pairs of legs and two pairs of long antennae. During a recent locust outbreak in Australia, the government launched a campaign aiming to introduce these insects into local cuisines. To help break popular reluctance to eat locusts the Australian media started to refer to them as "sky prawns." A similarly practical approach has been adopted by some people in Israel (eating locusts is allowed, with certain restrictions, by the Torah) and the Canary Islands.

In terms of their long-term survival, grasshoppers are no different from other organisms. Just as the white-tailed deer is a pest in some areas of the United States while its Asian relative the tamaraw (*Bubalus mindorensis*) is on its way to extinction, we battle locusts while critically endangered grasshoppers are disappearing. Several species, including the Central Valley grasshopper (*Conozoa hyalina*), have already vanished. In our conservation work on invertebrates we must not allow prejudices formed by a negative experience with some members of this group to influence our decisions to help other species. Habitat conservation is the best way to protect endangered grasshoppers and most other invertebrates. In addition, replacing indiscriminate insecticide spraying with more specific, narrowly targeted ways of controlling locust outbreaks is an important way to reduce damage to other harmless species.

Few places on Earth have a more species-rich fauna of grasshoppers than Australia. Dubbed "grasshopper country," Australia has been the center of a spectacular adaptive radiation of spur-throated grasshoppers (Catantopinae), and most species on this island continent belong to this group. They have filled niches usually occupied by other groups of grasshoppers, resulting in a remarkable morphological and behavioral convergence with various unrelated taxa, such as African stone grasshoppers (Pamphagidae). Others, like members of the genera *Stropis* and the aptly named *Zebratula,* resemble in their coloration and body form African bush hoppers (Pyrgomorphidae). The diversity of spur-throated grasshoppers is so high in Australia that despite being one of the best-studied regions of the world in terms of its insect fauna, hundreds of newly discovered genera and species of grasshoppers still await their formal description.

Common lagoonia *(Lagoonia scabronotum)*

Leopard grasshopper *(Stropis maculosa)*

Variable stropis *(Stropis nigrovitellina)*   Zebra grasshopper *(Zebratula flavonigra)*

Leopard grasshopper *(Stropis maculosa)*   Green stropis *(Stropis glaucescens)*

203

One must wonder whether the similarity between the pointillist style of Australian aboriginal paintings and the color pattern found in common Australian spotted pyrgomorph grasshoppers *(Monistria)* is accidental, or a genuine case of art imitating nature. The resemblance was pointed out to me by David Rentz, an entomologist extraordinaire who knows Australian grasshoppers like no one else on the planet. Aboriginal artists have been aware of grasshoppers for thousands of years, and they are featured in some cave paintings. Leichhardt's grasshoppers, close relatives of the spotted pyrgomorph, are a part of the aboriginal mythology, considered to be children of Namarrgon—the "lightning man" responsible for violent electrical storms in northern Australia.

The reason for spotted pyrgomorphs' gaudy coloration lies in their unpalatability, acquired through a diet consisting of a wide range of toxic plants. These grasshoppers sequester and store the plants' secondary compounds and advertise this fact through the universal language of warning coloration. The combination of black and yellow effectively signals the noxious properties of an animal, and patterns based on these two colors have evolved independently in many unrelated groups.

Foam grasshopper (*Dictyophorus cuisinieri*)

Spiny katydid (*Enyaliopsis* sp.)

Many southern African members of the grasshopper family Pyrgomorphidae exhibit beautiful, candy-like color patterns. While visiting friends in Zimbabwe some years ago I was warned by my hosts not to eat any because they are poisonous. Apparently, every year a number of local children become ill or even die after eating gorgeously colored nymphs of the African bushoppers (*Phymateus*). Although I had no intention of eating any, I was immediately attracted to these insects. They are slow moving and usually occur in large aggregations. When picked up they emit an acrid smell, and if further molested they fan their brightly colored hind wings and exude droplets of hemolymph (the insects' blood) through spiracles on their thoraxes. The hemolymph of bushoppers is their primary defensive weapon. These insects feed on a variety of plants rich in secondary compounds, such as milkweed (*Asclepias),* and many species of the plant families Apocynaceae and Scrophulariaceae, all of which contain extremely toxic cardiac glycosides and aglycones. The toxic compounds are sequestered by the insects and act as a very effective defensive mechanism. A bird must only try to swallow one tiny nymph of a bushopper to learn a lesson that will last a lifetime—these slow, tasty-looking animals cause instant vomiting and an illness that can last a few days. Grasshoppers of the genera *Dictyophorus (left, above and middle)* and *Taphronota* go a step further in their defensive behavior. In addition to ejecting their own toxic blood, they blow air into it through their thoracic spiracles, creating a layer of foam that gives a predator a taste of noxious chemicals before it even reaches the insect.

Some African katydids, like *Enyaliopsis (below left)* and related genera, also spew yellow droplets of their hemolymph if molested by a bird or entomologist, but they do not appear to be nearly as toxic as the grasshoppers.

Striped grasshopper (*Stenoscepa* sp.)

Variegated grasshopper (*Zonocerus variegatus*)

Green milkweed locust
(*Phymateus viridipes*)

Savanna grasshopper (*Heteracris* sp.) (Guinea)

Emerald grasshopper (*Kraussaria* sp.) (Guinea)

Savanna grasshopper (*Heteracris* sp.) (Guinea)

Being ignorant about life under our feet is a characteristic we share with many large, lumbering mammals. An area covered by a single elephant footprint in the dry savanna of the Okavango Delta is an arena of life and death drama, bustling cleanup activity, and efficient harvesting. Twigs and grass stalks crushed by elephants are collected by termites *(Hodotermes mossambicus)* and carried to an extensive network of underground nests. Grasshoppers such as *Pnorisa squalus* also feed on pieces of plants, but this innocent activity carries the risk of being captured and devoured by a grass mantis *(Hoplocorypha macra)*. These mantids are excellent mimics of dry stalks, and when prowling the sand they move in a way resembling a stick waving in the breeze. Mantids generally discard body parts that contain few nutrients but a lot of chitin, such as wings or legs. But these parts are not going to waste. They are carefully collected by many ants, such as *Pheidole* sp. and *Messor* sp., so long as these ants can avoid falling into sand pits dug out by voracious ant lions (Myrmeleontidae).

Grass mantid *(Hoplocorypha macra)*

Harvester termite *(Hodotermes mossambicus)*

Grasshopper *(Pnorisa squalus)*

A harvester ant *(Pheidole sp.)*

A hapless ant caught by an ant lion larva

If it were not for scarab beetles (Scarabaeidae), the African savannas would soon disappear under a thick layer of dung, copiously produced by elephants and other mammalian herbivores. Scarab or dung beetles are responsible for recycling a large proportion of animal waste, playing a double role of physically removing the large masses of material that could easily choke the vegetation if left on the plains and helping decompose it, returning the nutrients trapped in dung to the soil. The role of dung beetles is particularly well understood and appreciated in Australia. In the early 1960s Australian farmers realized that if something was not done quickly, the entire cattle industry in this country would collapse. The amount of dung produced by cows threatened to destroy most of this country's pastures. Cows are not native in Australia, and few species of Australian dung beetles are capable of feeding on cow dung. The enormous amounts of unprocessed dung also turned into a fertile breeding ground for bush flies and caused subsequent outbreaks of diseases carried by these insects. In a bold move, the Australian government decided to introduce a number of African, European, and North American species of dung beetles to Australia to help deal with the problem. About fifty species of dung beetles were introduced, and in the end the disaster was averted.

Euoniticellus kawanus
Oniticellus formosus

Onthophagus carbonarius

Among the introduced species was the bronze dung beetle *(Onitis alexis)* *(right)*, a common African species. Unlike the famous Egyptian sacred scarab *(Scarabaeus)*, females of this species do not roll balls of dung, but instead dig long tunnels directly under the pile of dung. At the end of the tunnel, the female packs a sausage-shaped mass of dung and lays her eggs in it. The larvae share the dung, each feeding on its own section, and once the dung is consumed they pupate in protective cases neatly constructed of their own feces. Although the biology of these beetles may appear unappetizing to us, their thankless activity keeps a lot of people fed and tropical savannas brimming with plant and animal life.

Savannas of Central America are home to a remarkable relationship between bullhorn *Acacia* trees and ants of the genus *Pseudomyrmex*. A long coexistence between these two organisms has produced a system in which the plant depends on the insects for protection from herbivores and other plants competing for the same space while the ants are incapable of living anywhere else. The ants regularly patrol the tree, warding off or killing any insect trying to feed on the leaves of the *Acacia*. In some cases they make an exception, allowing certain scale insects to feed on the branches—a behavior that clearly does not benefit the host plant. But they also remove seedlings of other plants from around the trunk of the *Acacia*, eliminating the possibility of the host tree becoming over-shadowed by others. They will even prune branches of neighboring trees if they threaten to overshadow the host. In fact, *Acacia* trees do very poorly and are easily outcompeted by other plants if the ants are absent. In return, the ants reap a number of benefits. At the base of the petiole of each *Acacia* leaf are small, elevated organs known as extrafloral nectaries that regularly exude droplets of fluid rich in sugar and amino acids. These are eagerly collected by the ants *(opposite, inset)*. Young leaves of *Acacia* also produce Beltian bodies—nutritious, protein- and lipid-rich packets that the ants use to feed their larvae. The tree itself is a home for the ants. Large, sharp spines on its branches are hollow, making perfect chambers to raise the brood and provide shelter for the queen and workers.

214

Pseudomyrmex flavicornis and *P. nigrocinctus* collect Beltian bodies from young leaves of *Acacia* to feed their larvae.

A worker of *Pseudomyrmex flavicornis* enters a hollowed-out spine of bullhorn *Acacia* that serves as a breeding chamber and a shelter for these ants. Young spines are solid inside and only become hollow as they age. The ants' only job is to chew out an opening to this ready-made home (Costa Rica, Guanacaste).

Perfectly concealed on a trunk of a savanna tree, an owl fly larva (Ascalaphidae) patiently waits for an unlucky insect to approach its long, sharp mandibles *(above)*. Like its close relatives, the ant lions (Myrmeleontidae), larvae of owl flies are missing two elements of the body that seem to be absolutely critical for any predator: the mouth and the anus. And yet they successfully survive their entire larval period without them. Their long mandibles are equipped with deep grooves that direct the liquid content of the body of their victims directly into their throat (pharynx). As for the lack of the anus, the larva's equivalent of kidneys, the Malpighian tubes, removes all waste products and stores them in the hind intestine. Once the insect reaches adulthood, the entire waste product of its larval life will be expelled in a single, highly concentrated white pellet.

Adults of owl flies and ant lions are also predaceous and resemble dragonflies in both their overall appearance and their hunting habits *(opposite)*. But like other members of the insect order Neuroptera, they are more closely related to beetles than dragonflies.

The most unusual members of the order Neuroptera are the mantidflies (Mantispidae). Adults of these insects, though unrelated, resemble preying mantids to an extraordinary degree *(below)*. Their neck region (pronotum) is strongly elongated and their front legs have evolved into powerful, grasping organs. Like true mantids, mantidflies feed on small insects, swiftly capturing them with their raptorial legs. The life cycle of these insects is as unusual as their appearance. Females lay very large clutches of tiny eggs, each egg attached to a leaf with a short, thick stalk. The minute, agile larvae that emerge shortly after quickly disperse in all directions looking for their only acceptable larval food—an egg sack of a spider. Finding such an elusive meal is not easy, and most young larvae will perish looking for it. Upon finding a silky cocoon full of spider eggs, the larva chews through its wall and undergoes a transformation into a grub-like, almost immobile stage. It then starts feeding on eggs and young spiders, eventually devouring the entire clutch. Once the food is gone, the mantidfly larva pupates and, after a while, leaves the now-empty egg sack as an adult.

Deep grooves and bare patches on the trunk of the baobab tree *(Adansonia digitata)* on the edge of the Okavango Delta in Botswana mark the favorite spots of elephants, who use the rough surface of the tree to rub their skin in order to remove ticks and other parasites. Restless young males also often puncture and rip the bark off with their tusks. This precarious environment is home to a surprising variety of life. In one night I found on a single baobab tree over fifty species of animals, including two species of lizards who found the perfect shelter under its loosened bark.

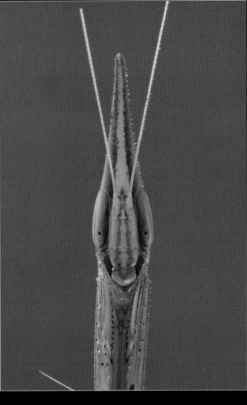

Many preying mantids common in African savannas have long, light brown bodies that perfectly mimic dry stems of grass. Their heads or even eyes have elongated conical shapes that complete the illusion. But savanna mantids often exhibit fire melanism, in which black individuals predominate in a population following a grassland fire.

Some species of savanna mantids hunt insects from grass stems by employing the sit-and-wait strategy, while others actively patrol exposed patches of sand. When they walk, their movements resemble a dead piece of grass being blown by the wind.

Cone-headed stem mantid (*Pyrgomantis* sp.) (Guinea)

Double-coned grass mantid *(Episcopomantis chalybea)* (Botswana)

Grass mantid (*Hoplocorypha* sp.) (Botswana)

African savanna mantids *Polyspilota aeruginosa* and *Miomantis* sp. (Guinea)

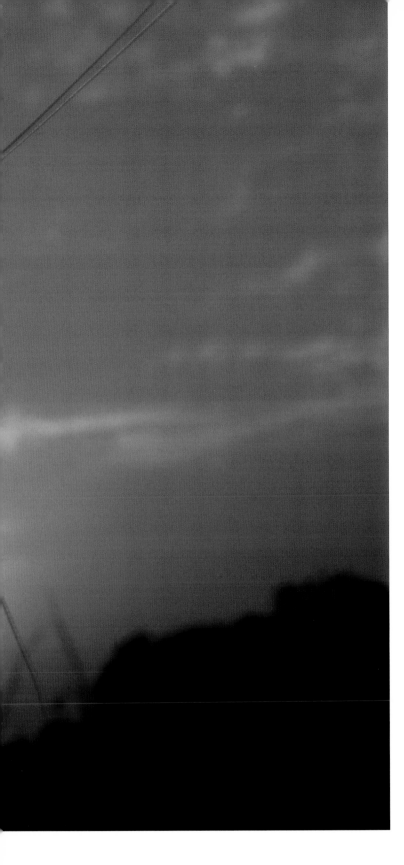

As the sun sets over Botswana's Okavango Delta, reed frogs wake up. Fragile appearance notwithstanding, frogs of the genus *Hyperolius* are champions of survival in the harsh conditions of the southern African savanna. Their skin contains chromatophores—special cells that by contracting or expanding can change body color from snow white to darkly patterned. White coloration is possible thanks to the presence of guanine and hypoxanthine crystals in cells known as iridophores. It is assumed by reed frogs during the day, effectively reflecting rays of the sun and lowering their body temperature and thus the loss of water. Their skin also retains several layers of epidermis, normally shed by other frogs, which in combination with hardening mucus, engulf the frog in a thin yet nearly impenetrable protective pocket.

Some species such as *Hyperolius marmoratus (above left and right)* exhibit remarkable variation in their color patterns, and no two individuals are alike *(right; opposite, above)*. Other species, like *Hyperolius nasutus (above; opposite, below),* are more conservative, donning plain green patterns that make them disappear on stems of papyrus, where they spend their days.

DESERTS

THE FIRST NIGHT I SPENT in a tent in the Kalahari was also one of the coldest in my life. Even though I knew that the temperature differences between night and day could be significant in a desert, nothing had prepared me for the near-freezing experience that followed the blistering heat of the day before. Putting on a second pair of pants and another layer of T-shirts, I cursed myself for not being better prepared. I learned my lesson that night, and on my next trip to the desert I stocked up on warm clothes. Unfortunately, my luggage had been lost on its way from Johannesburg to Windhoek, and I only saw it again after spending three weeks shivering each night in a borrowed, thin sleeping bag, with just a single shirt on my back.

The daily difference in temperature in desert ecosystems, which regularly approaches 40°C (~70°F), is rooted in the extremely low humidity of these environments. Deserts lack the buffering effect normally brought by water in the atmosphere. The air above the desert is exceptionally clear, and the lack of water in it means that nothing intercepts and diffuses solar radiation before it reaches the surface. The ground re-radiates the heat, and because there is nothing to trap it, the ground continues to cool throughout the night. This nocturnal cooling is probably the reason why many desert animals, such as ants or darkling beetles, are black, despite the fact that black coloration may cause overheating. Apparently, the ability to absorb solar radiation quickly in the morning and warm up enough to start foraging early is worth the trouble of dealing with excessive heat absorption later during the day. Of course, it would be ideal if an animal could change its color from black to white, depending on its need to warm up or cool off. Although such a thermoregulatory trick is impossible for animals with hard, external exoskeletons, it is entirely within the realm of possibility for reptiles, and the Namaqua chameleon and many geckos can change their coloration from black to almost snow white thanks to numerous chromatophores in their skin.

But high temperatures, which can easily reach 75°C (167°F) on the surface of the sand are not the only problem desert animals must deal with. Solar radiation itself, high in harmful ultraviolet (UV) waves, can increase the rate of genetic mutations, generally a very negative phenomenon. Animals deal with this problem in three different ways. The most obvious approach to avoiding excessive solar radiation is to hide from it. Many small animals, such as beetles, silverfish (Thysanura), some crickets, and lizards spend most of their active time under the surface of the sand; others hide under rocks or in thick clumps of grass and other vegetation. Scorpions and solifuges limit the amount of UV radiation penetrating their bodies by including on the surface of their cuticle compounds that effectively reflect these wavelengths. A different strategy has evolved in most desert lizards, whose peritoneums, the smooth membranes that line the body cavity, contain dark pigmentation. This pigmentation absorbs most UV waves before they reach vital organs. It is possible that the dark coloration seen in many desert organisms, in addition to helping them raise their body temperature in the morning, also acts as an absorbent barrier to UV radiation.

Desert surfaces vary depending on their geologic history, ranging from hard, rocky terrain to the fine, almost liquid-like sands we typically associate with this environment. Walking on loose sand can be exhausting, as anyone who has ever walked on a sandy beach can attest. Desert animals have evolved a number of adaptations to help them solve the problem of locomotion on a shifting substrate. Some of the solutions are remarkably similar to those found in aquatic and semi-aquatic organisms. Many lizards have fringed fingers much like those found in semi-aquatic lizards such as basilisks. Small flaps of skin on each digit increase the surface area of the foot, preventing it from sinking into the sand. In an astounding example of convergent evolution some desert lizards have developed bipedal locomotion, seen also in lizards capable of running on the surface of water. The gecko *Palmatogecko rangeri* complements the aquatic analogy by having webbed feet that allow it to run on very fine, loose sand. The webbing is supported by a system of cartilages, which combined with a unique musculature allow the gecko to

use its feet to scoop sand when digging burrows. Splay-footed crickets *(Comicus)* deal with the problem of locomotion on loose sand in a similar fashion, having evolved long, lateral extensions on their hind feet. Some organisms forego trying to stay on the surface of the sand and instead "swim" within it. A number of species of skinks are capable of moving in the sand using undulating motions that resemble those of a swimming snake. They are capable of extracting air from between grains of sand and only rarely surface.

The trials of life in the desert do not end with high temperatures, radiation, or the difficulty of staying on shifting surfaces. Unquestionably, the greatest predicament of all desert organisms is the scarcity of water. Coastal deserts such as the Namib or Atacama receive most of their moisture in the form of an early morning fog from the ocean and may not see actual rain for many years. Rains are rare and often unpredictable in most other deserts, and high temperatures increase water loss from the animals' bodies. Thus, animals must find and retain water. Large mammals and birds regularly migrate to known sources of water, but smaller animals must find other ways. Nocturnal temperature drops have the positive effect of causing the minute amounts of water found in the desert air to condense, and many animals rely on this source of the precious commodity. Long-legged species of darkling beetles (Tenebrionidae) employ a behavior known as "fog basking," in which they stand on the top of desert dunes early in the morning, allowing tiny droplets of moisture to accumulate on their bodies. By tilting their bodies forward, they cause the beads of dew to aggregate and form larger drops that flow toward their mouthparts. Beetles of the southern African genus *Lepidochora*, having short legs and a flat body that preclude them from using the "fog basking" technique, build small trenches on the seaward slopes of dunes that effectively trap water, allowing the insects to lick it off grains of sand. In the ultimate water-trapping behavior, larvae of some darkling beetles have evolved the ability to capture water directly from the humid morning air by using high osmotic gradients in a specialized area of their rectums.

Water retention is also a major problem, and small animals address this issue in many ways. The shovel-snouted lizard *(Melores anchietae)* has a specialized water-storing bladder along its lower intestine, and the animal can draw on this reserve for several weeks. But most animals cannot rely on such a luxury and must minimize water loss by any means possible. Breathing, the unavoidable necessity of all animals, is a major source of water loss. Some insects, including the Namib dune ant *(Camponotus detritus)*, employ a behavior known as discontinuous ventilation cycling. Instead of taking regular "breaths" through their spiracles, they keep them tightly sealed for as long as several hours and exchange gases in short, irregular bursts. Darkling beetles reduce respiratory water loss even further by having their spiracles open dorsally under their thick, permanently sealed elytra. Other insects produce waxes on the surface of their cuticles that reduce evaporation.

Considering all the hardships of living in the desert, it may come as a surprise that 15 percent of the entire human population, or nearly a billion people, make their homes in the desert environment. Such an enormous number of human inhabitants makes a significant impact on the arid areas of the world.

Although not all that happens in the desert as a result of human presence is bad, there are human activities that cause a lot of harm. Mining operations and cattle grazing are the two most serious causes of desert habitat degradation, followed by the introduction of nonnative plant species. In some deserts unchecked use of off-road vehicles destroys fragile plant cover and unique lichen communities known as crust, which can take decades to recover.

As is always the case, there are no easy answers for desert conservation. People who already live there must be able to use available desert resources, and mining, which in some cases is the main source of income for desert nations, will go on for a long time into the future. Establishing restricted access zones around particularly unique or threatened arid areas is definitely the best solution, but the enforcement of their protected status is often problematic. Although there is no risk of deserts ever disappearing from the surface of our planet, as we fear rainforests may, preservation of the natural status of these exceptional ecosystems should be considered as important as maintaining the remainder of Earth's biota.

Similar conditions in deserts on different continents result in the evolution of similar adaptations to life in these extremely dry environments. Most insect species have fringed or hairy legs that increase their surface area and allow them to walk on fine sands without sinking. A large proportion of insects found in sandy deserts can also bury themselves in the sand in the blink of an eye, including such unlikely diggers as the Australian grasshopper (*Urnisiella*) (*below*).

The seemingly lifeless, constantly
shifting dunes of the Namib desert
are home to a surprisingly diverse fauna.
Most animals living there are exceptionally
well adapted to dealing with both the blistering
heat of the day and the loose, unstable sand.

The Smith's desert lizard
(Meroles ctenodactylus)
has a number of adapta-
tions to life in the fine,
loose substrate of the
dunes: a wedge-shaped
snout to help it dive into
the sand, fringed fingers
to increase the surface
area of its feet, and a skin
fold that covers the ears
to protect them from
being clogged with sand.

A darkling beetle climbs a dune in search of plant and animal debris blown in by winds from other areas of the desert.

When cornered, this giant ground gecko (*Chondrodactylus angulifer*) will hiss loudly and arch its stubby tail over its back in an uncanny imitation of a scorpion. For this reason, despite being absolutely harmless, this species is feared by people across its range in Namibia and South Africa.

At home in one of the harshest environments on the planet, the Namaqua chameleon *(Chamaeleo namaquensis)* stalks insects, lizards, and small snakes on scorching hot sands of the desert alongside the desolate Skeleton Coast of Namibia. A nasal salt gland allows it to forage close to the Namib desert's seashore by removing excess salt from its food and water. Namaqua chameleons are almost exclusively terrestrial, although their clasping feet and a prehensile tail bear witness to a relatively recent departure from the arboreal lifestyle typical of most chameleons.

The ability to change its body color helps this chameleon to effectively regulate its body temperature. Desert nights can be very cold, and so the animal starts the day by darkening its skin and exposing it to the rising sun to warm up quickly and start hunting as soon as possible. As the sun rises over the horizon and the sand becomes increasingly hotter, the chameleon becomes lighter, often almost white, to help deflect the sun's rays and lower its body temperature.

Perched atop a welwitchia *(Welwitchia mirabilis)* early in the morning, a Namaqua chameleon inflates its body and darkens its skin to warm up quickly in the rays of the rising sun (Namibia, Namib desert).

Life takes hold even on the blistering hot red granite rocks of Namibia. Vascular plants would be hard pressed to survive in this environment, which is nearly completely devoid of water, but lichens thrive here in a surprising variety. Their presence allows many species of grass-hoppers to live here as well, such as the two genera of rockhoppers, *Lithidium* and *Lihidiopsis*, found only in the hottest and driest areas of southern Africa. Their wingless bodies look as if they were constructed of tiny pieces of granite, and their cryptic coloration and texture make them disappear on the fully exposed, naked rock. And hide they should. There is not much else available for the Boulton's day gecko *(Rhoptropus boultoni)* and Namibian rock agama *(Agama planiceps)* to feed on in this extremely harsh environment.

Male *Agama planiceps*

Boulton's Namib day gecko *(Rhoptropus boultoni)*

239

A female of *Uroplectes plani-manus* will carry her newly born offspring on her back until their cuticles are hard enough to protect them from desiccation and they are ready to hunt on their own.

A pair of comb-like organs (pectines) on the ventral surface of their bodies allows scorpions such as *Opistophthalmus carinatus* to detect even minute traces of moisture, although most of their water intake comes from their insect prey.

Protected by their cuticles, which effectively reflect ultraviolet light,
scorpions can survive in the scorching heat of the Namib desert.

Solifuges, or sun spiders, are the invertebrate equivalents of mammalian solitary predators such as the African badger or Tasmanian devil. They seem to have boundless energy that allows them to run for hours, actively searching for prey, which may include anything from a cricket to a small lizard. Solifuges do not have venom glands, but their enormous mouthparts, or chelicerae, are strong enough to kill prey at least as big as themselves. Because sun spiders cannot ingest large pieces, the prey is injected with digestive enzymes and masticated to a pulp before swallowing.

Their common name is misleading. Most of these animals are exclusively nocturnal and may die of heat stroke if exposed to direct sunlight. Their Latin name, Solifugae, meaning "fleeing from the sun," seems more apt.

African countries, these animals are often feared because some diurnal species appear to chase people. In fact, they are simply following the shadow created by a person, trying to avoid direct exposure to the sun's heat. They are also accused of cutting patches of hair from sleeping animals and people, resulting in another common name, the haircutters. Interestingly, clumps of mammal hair have been found in nests of some sun spiders, indicating that females may indeed use it, but whether the hair is cut or simply collected from the ground is unknown.

A soft whisper of grains of sand pushed by the desert wind is the sound usually associated with the vast expanses of the African deserts. But even there, hiding in the sparse, shrubby vegetation of the Namib, a number of singers interrupt the prevailing silence with their love serenades. Some of them make calls like no other animal on Earth. The huge grasshoppers of the genus *Lamarckiana* (Pamphagidae) produce a call so strange that when I first heard it one evening in Namibia, I was absolutely convinced that an alien spaceship had landed in the nearby hills. Males of these insects have long, leathery wings, the central portion of which is somewhat depressed and covered with tightly spaced, parallel ridges. Rather than rub their femora against the wings, as any "normal" grasshopper would do, they rapidly kick their feet against the ridges, resulting in a unique sound that combines the swish of a rocket engine with the loud clatter of a stick dragged against a wooden picket fence. Females of *Lamarckiana*, named so in honor of Jean-Baptiste Lamarck, the famous French zoologist and proto-evolutionist, are completely wingless but also produce loud, rattling noises by rubbing their legs against the rough surface of their abdomens.

Of the few katydid species that survive in the arid environment of the Namib, the mottled *Pseudosaga maculata (below)* makes probably the loudest continuous buzzing call, whereas the large and fiercely predaceous long-winged clonia *(Clonia caudata) (left)* is a katydid with a surprisingly soft and short call.

Isolated by the Namib, one of the oldest deserts in the world, stands a massive volcanic plug known as Massif Brandberg. In 2002 its plateau was the location of a discovery that surprised entomologists around the world—an entirely new order of insects, the heelwalkers or Mantophasmatodea, was found to live among its rocky outcrops. Even more surprisingly, this group of insects was thought to be extinct for nearly 50 million years, and finding them alive today is akin to running across a living mastodon. As I walked among the red rocks of Brandberg as a member of the first expedition to collect live heelwalkers, I felt the elation that only comes from being in the presence of something new and absolutely unique. But to an unprepared eye, these inch-long, wingless gray or green insects may not seem particularly interesting. In fact, they have been collected by entomologists for over a hundred years and many natural history museums have them in their collections. And yet during this time not a single scientist gave them more than a cursory glance before dismissing them as

larval stages of some other common insects. It was not until Oliver Zompro, a German student of entomology, realized that insects identical to those he had seen in ancient amber were among recently collected specimens from Namibia that heelwalkers were properly recognized as a distinct insect order.

Heelwalkers are sit-and-wait predators, somewhat similar in their behavior and appearance to small preying mantids. But their affinities lie with another obscure, relict group of insects, the ice crawlers (Grylloblattodea), found in the snows of North America and China. Some heelwalkers are active at night, feeding on small insects and spiders, while others hunt during the day. Their common name is derived from the peculiar way their feet are held while walking. Since their discovery, more than ten additional species of heelwalkers have been found in South Africa, Namibia, and Tanzania, and more are likely to be found in dry habitats of Africa.

Wingless and larval in appearance, it is not surprising that heelwalkers, like this adult female of *Mantophasma zephyra*, have for many years escaped the attention of entomologists.

Heelwalker *M. zephyra* is a diurnal,
sit-and-wait predator.

Heelwalkers lay their eggs in the sand, encased in
a foamy mass that protects them from desiccation.

The gladiator heelwalker *(Tyrannophasma gladiator)* is a ferocious nocturnal predator.
It uses both front and middle legs to grab and crush its spider and insect prey.

The beauty of the tropics is that no matter how long you work there, something will always surprise you. I found this new genus and species of leaf katydid after ten years of intensive sampling at La Selva Biological Station in Costa Rica.

Epilogue

New Life

WHEN I WAS ABOUT SEVEN YEARS OLD my father said something that, in retrospect, was probably the most important piece of information I have ever received, one that has influenced my entire life. I seem to recall that we were talking about the recent landing on the moon by the crew of Apollo 17. I should mention that my father was an astronomer, and he did his best to try to entice me with his line of work, but to no avail. To my question about the possibility of finding life on the moon he replied that those chances were very slim, but that our own planet still had many organisms that remained undiscovered and unnamed. That statement stopped me in my tracks. I instantly envisioned mysterious, dinosaur-like creatures hiding in the depths of African jungles. But he explained that the yet-unnamed animals were probably small, and that they might even be living in our own backyard. I am certain that at that very moment I decided to become an explorer who would spend his life questing for those elusive, undiscovered creatures. Of course, I did not realize then that most discoveries of new species take place among dusty drawers of museums, made by taxonomists whose lives are decidedly less than glamorous. But the stage for a career in entomology was set.

It took me fifteen years to run into a beautiful little gem of a katydid in a meadow in Turkey, an animal that until then had eluded entomologists. I named it *Poecilimon marmaraensis*, after the Marmara Sea on the shores of which this species occurs. The elation of knowing that I was the first person on Earth to find it was intoxicating. I am sure that all my fellow taxonomists have felt it at some point in their lives, and for many it is the ultimate reward. Finding new forms of life is to me one of the greatest adventures in biology, one that the current biodiversity crisis makes a pressing necessity.

Extinctions and the appearance of new species are natural phenomena that once were fairly well balanced, with speciation always slightly outpacing extinction, making Earth's biodiversity increasingly richer. With the advent of man, this balance started to tip dramatically toward extinction, and we now lose thousands of species every year. In most cases we do not know for sure what species are gone because our knowledge of many groups of organisms is rudimentary. All we can do is to produce estimates of extinction based on our understanding of population sizes, dispersal abilities, and environmental tolerance of related taxa. In only comparatively few cases can we authoritatively say that a particular, named species has been wiped out.

Regardless of being named or not, each species gone extinct is another book of genetic code burned, another piece of the intricate puzzle of life on Earth irreversibly lost. In light of the recent revival of plans for space exploration with the explicit goal of searching for extraterrestrial life, it never ceases to amaze me how little interest there is in investigating organismal aspects of our own planet. At the same time, most of the scientific community agrees that as the twentieth century belonged to technology, the present century will belong to biology. This assumption is based not only on the accelerating rate of discovery in biomedical and biotechnological fields, but also on the belief in our ability to deal with large amounts of data. And while destruction of natural habitats and the alarming pace of species extinction leaves little hope for preserving all the components of Earth's biota for future generations, we should make every effort to document all organisms that still inhabit our planet. Armed with this knowledge, we will be able to make more informed decisions as to where conservation efforts should be focused in order to preserve the highest proportion of evolutionary diversity and biocomplexity. The more diverse the life, the more likely we are to discover within it new medicines, new crops, or new structures worth imitating. This principle, that in order to see the big picture you must have the complete catalog of all its components, is well understood by the biomedical community. It was clear that significant advances in human genomics could not be made until the entire genome was mapped, even if the process of obtaining the map was costly, time-consuming, and without immediate practical benefits. Now it is time to do the same for our biodiverse planet.

In order for such an effort to succeed, a significant cultural shift must occur within the biological community with respect to dealing with taxonomic data. Basic, descriptive work needs to return to the toolbox of modern biologists, this time augmented with a better understanding of evolutionary processes and the many technological advances that have occurred since humans first started to classify living organisms.

A sentiment sometimes heard among researchers that descriptive, taxonomic work is not a true science is an expression of ignorance, forgetfulness that virtually all great biological principles and discoveries originated from countless individual observations. There really is no conceptual difference between describing individual species of cockroaches in order to get the complete picture of the evolution of protozoan symbiosis in their close relatives, termites, and sequencing thousands of fragments of DNA in order to get a base-by-base map of the entire human genome. Both processes involve collecting data points (species or bases), and both can result in the ability to ask and answer questions that may have not been envisioned when the process was started.

How much life is there still to discover on our unique, green planet? Quite honestly, nobody knows for sure. In fact, nobody knows exactly how many species have already been discovered and described, although the consensus seems to oscillate around the figure of 1.7 million species. Estimates of the total species richness vary from a conservative 3–4 million to over 100 million species, and the truth lies probably somewhere around 5–10 million species still to be found. Most of these yet unknown organisms are probably insects and nematodes, but not all new species are small invertebrates. Species of frogs, lizards, and even birds and monkeys previously unknown to science are frequently being found in remote forests of the tropics. Even such unlikely places as metropolitan areas of developed nations can occasionally surprise scientists with a new centipede or salamander.

In my work on insects and arachnids I have so far discovered about eighty new species, some unique enough to deserve placement in a higher, separate taxonomic category, a new genus. But my findings pale in comparison with those of the most prolific taxonomists, some of whom have found and described thousands of new taxa. The following photographs show a few of the newly discovered, still unnamed animals my colleagues and I have found. I hope that some readers will feel the thrill of knowing that few before them have ever seen these elusive neighbors of ours.

Traveling in Australia with David Rentz, an entomologist, was like having an exclusive, personalized tour of a lost treasure trove of biodiversity. In the last thirty years David single-handedly increased the number of species of grasshoppers and katydids known in Australia from about 200 to several thousands. Many of them still await a formal description, including several species common in his own garden in Queensland.

Males of this still-unnamed genus and species of grasshopper are wingless and thus incapable of producing songs. Instead, to attract females they use their brightly colored hind legs in a precisely choreographed courtship display.

A new genus and species of katydid from David's garden in Kuranda, Queensland

Another new species of katydid from Kuranda

253

A new species of
the moss-mimicking
katydid genus
Paraphidnia
(Costa Rica)

Costa Rica is probably the tropical country most thoroughly explored by zoologists and botanists in the world, but new species are still being discovered almost daily. One of the most surprising findings was an aquatic cockroach, seen here with a nymph of a damselfly in a stream in the Braulio Carrillo National Park.

My own specialty, the katydids, also do not disappoint. Nearly 30 percent of all katydid species in Costa Rica are new to science. This yet-unnamed leaf mimic lives in the cloud forest of Braulio Carrillo National Park (Costa Rica).

Of all the tropical regions of the world, West and Central Africa are probably the least biologically explored. Political instability and frequent violence hinder attempts to establish long-term research programs there, and a multitude of dangerous parasites and diseases for which this area is famous makes a biologist's work even more difficult. But despite these obstacles, biologists flock to this area, attracted by the prospect of new discoveries and the need to catalog its life before uninhibited logging and mining destroys most of it. Sometimes the only way to collect precious biological data in parts of the world that are difficult to access is to assemble a team of specialists and like a military commando descend on a threatened forest or a mangrove swamp to quickly collect as much information about organisms living there as possible. The Rapid Assessment Program within the nonprofit organization Conservation International does exactly this, and the results are often quite remarkable. I had the opportunity to join a few such expeditions, including one to the highly threatened afromontane savannas in the Simandou Range of Guinea. Within just a few days, our team discovered a wide variety of organisms new to science, including yet unnamed frogs, katydids, and cockroaches as well as a variety of rare and endangered birds and mammals.

I found this gorgeous, predaceous katydid belonging to an unnamed species so distinct that it will also be placed in a separate, new genus.

This unnamed species of the frog genus *Arthroleptis* was discovered by Mark-Oliver Rödel, a herpetologist.

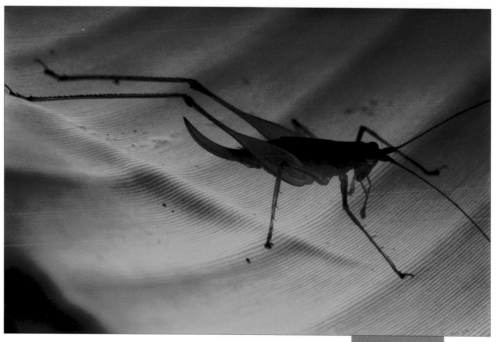

A newly discovered katydid species, spotted along a stream in the high-elevation savanna of Mt. Pic du Fon in Guinea.

Photographing the Smaller Majority

URING THE LAST TEN YEARS, I have spent a considerable
amount of time working as an entomologist in various
tropical destinations on projects involving biology and conservation of insects and other invertebrates. What started as an attempt to simply document some of these organisms with a camera
my wife had surprised me with one Christmas day quickly developed
into a photographic passion for capturing all things small and over-
looked by many nature photographers. Very early on I discovered
that it gives me immensely more satisfaction to lower my lens and
look for animals hiding on the forest floor than to take a picture of
an elephant or toucan, subjects that have already been expertly photographed. I also realized that by focusing my attention on smaller
organisms I had gained access to a hidden, unexplored treasure trove
of nature, full of exquisite gems of colors, shapes, and unique behaviors. To understand my fascination with smaller animals try this
simple exercise: imagine a herd of elephants, but one in which each
animal is only the size of your fingernail. Imagine looking at them
from above as they walk around your feet. Pretty unremarkable gray
specks, wouldn't you agree? Now imagine a preying mantis the size

of an actual elephant—isn't it the most spectacular creature you have ever seen? My point is that once you remove the element of size, you will start noticing the overwhelming variety and beauty of small animals that share this planet with us, revealing a diversity of forms most people do not realize exists.

Capturing this smaller majority of life on film, while not without its challenges, is not as difficult as you might think. Taking pictures of minute organisms is really not that different from other types of photography, and with the help of a few accessories you will be able to enter the exciting realm of tiger beetles, ant lions, and giraffe weevils. To find subjects for macrophotography you do not need to travel far. Your own garden is full of inhabitants you never knew were there, and a nearby meadow or forest will provide enough subjects to satisfy even the most de-

manding nature photographer. But it is the tropics where you will find some of the most amazing creatures that have ever walked the Earth, and if you have a chance to do so, I encourage you to take a photographic trip there.

Day One in the Rainforest

Of all habitats in the tropics, rainforests are probably the most challenging places for a nature photographer, but at the same time they may also be the most rewarding. First-time visitors frequently exhibit one of two types of reaction in their encounter with the rainforest environment. One group includes those who expect a zoo or a tropical paradise, and find themselves disappointed by the lack of multitudes of gorgeous flowers hanging from every branch or herds of animals lurking behind every tree. The image of the rainforest most of us have in mind has been shaped by years of natural history documentaries and books such as this one, which condense years of research and observations. In real life, chances are that you will see few flowers (although those that you will see may be spectacular), and larger animals will appear as small, fleeting shadows high in the canopy.

Visitors in the second group find themselves completely overwhelmed by the richness and diversity of the forest. These are the people who do not expect to see the few well-publicized rainforest icons, such as jaguars or orangutans, but instead are open to every single life form and experience of their trip. I remember well the sensory overload of my first visit to a Malaysian rainforest. The thing that struck me the most was how difficult it was to find two organisms of the same species, whether butterflies or trees. I wanted to take in everything, remember every little detail, and take a picture of every single organism. This sensory overload led to exhaustion, which left me feeling numb to the surrounding diversity.

Neither approach will result in spectacular photographic documentation of your visit. You must first acclimate yourself to the new environment and learn to be selective in choosing your subjects. For those visiting the rainforest for the first time, I have the following suggestion: on the first day of your visit to a tropical rainforest, leave your camera behind. Instead, take a slow walk along a trail, breathe in the ambience of the site, listen to its sounds. Look up and notice the variety of plant life above your head. Stop every now and then and take a closer look at leaves and tree bark. You probably will not see too many animals at first, but as you continue scanning the vegetation you will start noticing some of the inhabitants. Don't be afraid to turn over fallen twigs or gently unroll dried leaves. Although rainforests have their share of biting and stinging animals, by taking common sense precautions you will avoid unpleasant encounters. Do not make foolish mistakes such as sticking your arm into a tree hole without checking what lives in it. It may be a cobra's shelter, as I discovered the hard way (luckily, I wasn't bitten). Remember to leave everything in exactly the same condition as you found it. Don't break plants and don't strip bark off tree trunks. If you turn over a rock, gently put it back in exactly the same spot. Always keep in mind the scale of things: a small rock for you is an entire ecosystem for a colony of springtails and a nursery for a family of crickets.

The slower you walk and the more patient you are, the greater the rewards. The whole place will begin to unfold before your eyes. Not only will you start seeing beetles and sleeping frogs, you will also notice the gorgeous textures of plants (don't forget to look at the underside of leaves, too) and fantastic shapes of lichens and fungi, which in some places cover virtually every available surface. Look even more closely and you will discover miniature gardens of moss and tiny bromeliads on twigs and bark; if you are lucky a piece of bark may suddenly move, revealing its true identity—a camouflaged gecko.

After a while you will start forming ideas of what you would like to photograph and what technical difficulties you will have to overcome to do so successfully.

Rainforest Challenges

As you may already suspect, the number one enemy of the rainforest photographer is water. It is everywhere—suspended in the air, dripping from leaves, splashing under your feet, and pouring as a torrential rain down your neck every few hours or so. But none of this will matter if you are prepared. Bring plenty of plastic bags, and make sure at least

one of them is large enough to hold your camera with the lens attached. Pack every piece of equipment into a sealed bag and keep it there at all times, preferably with a packet of rechargeable silica gel to absorb any extra humidity. Unless your camera bag has a waterproof hood, carry a large plastic sheet with you (a large garbage bag will do) to cover all of your equipment in case of rain. This sheet will double as a waterproof platform that you can spread on the ground to safely assemble your camera or kneel on to take a ground-level shot of a marching column of leaf-cutter ants. I also sometimes carry a small umbrella, which comes in handy when I want to shoot during the rain. Remember that no matter what, you *will* get wet;

what is important is to keep your gear dry at all times. Incidentally, you will be faced with very similar problems when photographing in the desert or dry grasslands, except that instead of water your camera will be threatened by fine dust and sand. Plastic bags are also very effective to keep these elements away from your equipment.

There are many negative side effects of the high humidity typical of most rainforest locations. For one, any differences in the temperature between the camera and the air (e.g., if your camera has spent some time in an air-conditioned car or a hotel room) will result in water condensation on the camera body and lens, effectively preventing you from taking any pictures until their temperatures have equalized. It is a good idea to let your equipment (protected in plastic bags) sit outside for at least thirty minutes before starting to shoot. Even then, condensation will appear on the camera under your fingers, and sweat from your forehead will drip on the viewfinder. I always carry plenty of soft paper tissue with me to wipe water off my equipment. If available for your camera, consider installing a nonfogging eyepiece on the viewfinder. Water condensation has the magical property of appearing on the viewfinder a second before you are ready to take that shot of a preying mantid capturing a hummingbird, forcing you to wipe the glass and consequently scare away the animals. For the same reason, if, like me, you are forced to wear corrective lenses, consider wearing contact lenses instead of glasses while on your rainforest safari. Not only will water not condense on your lenses while your eye is pressed against the camera, but during the inevitable rain, you will not find yourself half-blinded by water streaking down your glasses.

Because of the typically low level of light in the forest interior, you have two choices when it comes to illuminating your subjects: using a flash or making long time exposures. The latter works for insects and other small animals such as katydids, caterpillars, or geckos that rely for their protection on being cryptically colored or shaped, as they tend to sit still and allow you to get really close. The near-complete lack of wind in the rainforest will allow for long time exposures without blurring your photographs, although you will have to use a tripod or some other method of immobilizing your camera while taking the picture. Upon finding an interesting and cooperative subject, I like to experiment with different focal lengths, first taking a few "documentary" shots with a 180 mm macro lens, and later switching to a wide-angle lens that allows me to compose more dramatic shots that incorporate not just the main subject but also a large portion of its habitat. Before you start taking pictures of an animal, spend a few minutes observing its behavior. See if it moves as you get closer, and if it does, note the pattern of its movement. Many small animals exhibit an interesting, albeit somewhat frustrating behavior of playing "peek-a-boo" with the photographer, always trying to be on the opposite side of the branch or tree trunk. If this happens, and your gorgeous tree mantid or lizard runs to the other side of the trunk, don't worry. Calmly assemble your equipment, compose the shot, and then gently reach and wave your hand behind the tree trunk. More often than not, the animal will come rushing back, right in front of your camera.

It goes without saying that long time exposures require sturdy camera support. A tripod is an obvious, albeit often heavy and bulky choice. Instead, you may consider a different solution. As many of my shots are taken at ground level, I find that a flat piece of board with a tripod head attached does a much better job than an unwieldy tripod. You can easily make one yourself out of an inch thick eight-by-ten-inch piece of wooden board. Simply drill a hole in the middle to screw on your tripod head and you have a more versatile support for low-level shots than any tripod could provide. Attach small rubber feet to it so it will not slip on most surfaces, and you can easily place it on a fallen log or a rock to get a slightly higher point of view. An added bonus is that any wide-angle shots taken from this low vantage point will appear more original and dramatic than the same shots taken from eye level. The only drawback of shooting from ground level is that unless you have a digital camera with a swiveling preview screen, you will have to either lie flat on your stomach to peek into the viewfinder or purchase an angle viewfinder, available for many single lens reflex (SLR) cameras.

The long exposure technique, however, will not work if you are hoping to photograph a marching column of ants, a fluttering butterfly, or any other type of fast behavior in the dim light of the forest floor. In such a situation, a flash is a must, and a good, dedicated TTL (through-the-lens light metering) flash will take all guesswork out of your macrophotography. There are as many opinions on what the ideal macro flash setup should be as there are photographers. Your decision should be based on the size of your intended subjects and the type of lens you will use. Many helpful books on macrophotographic techniques discuss the issue of the choice of flash in great detail. For very small subjects, a specialized "ring flash" may be a good choice. The common opinion among professional photographers is that this kind of illumination produces "flat" lighting. Although this is true for objects that are far away from the lens, for small objects, photographed from a distance equal to or only slightly greater than the diameter of your lens, a ring flash will produce photographs with well-defined shadows and sharp relief.

At Night

The most spectacular inhabitants of the rainforest can be seen only after dark. This is the time when most insects and other invertebrates as well as most amphibians and many reptiles start foraging, safe from the sharp eyes of birds and monkeys. Paradoxically, for a person armed with a flashlight, even the most cryptic species are easier to find and approach at night than during the day. Even larger animals, such as armadillos or nightjars, unprepared for the bright illumination of your portable light, will allow you to get close—impossible during the day. You may also see sleeping animals, including birds and butterflies. Nighttime macrophotography is not as difficult as it may sound, although you are limited to flash illumination.

Walking at night in the rainforest is no more dangerous than during the day. In fact, you are less likely

to step on a snake at night than during the day, as most of them are active and alert and will quickly move away, sensing your approach. Still, as a preventive measure I highly recommend wearing tall leather or rubber boots that reach at least above your ankles. Unfortunately, mosquitoes are also at the peak of their activity at night and in some tropical locations they carry malaria and other diseases. Wear a long-sleeved shirt and long pants and use insect repellent on exposed body parts. Having experienced typhoid fever and filariasis because of my negligence to apply insect repellant, I know how important it is. Be careful not to spray any of it on your photographic equipment. Some repellents can damage the plastic parts of the lens and the camera.

Make sure that you have a good source of light with you. A strong headlamp will free your hands. In addition to the main light source I always bring a small backup flashlight with me. I have been stranded in the absolute darkness of the rainforest more than once by a burned-out headlamp bulb, and it is not a pleasant experience. This extra light will also help you illuminate and focus on your subject.

If you are using an SLR camera when photographing at night, a shorter lens is preferred; 55mm or 60 mm macro lenses are ideal. You can also add a diopter lens or extension rings to increase the magnification factor. Longer focal lengths (100–200 mm) are useful when you want to isolate your subject from a well-illuminated background during the day, but at night your subjects will already be perfectly defined against the uniform blackness. A longer lens also requires a greater working distance, but at night you want to be as close to your subject as possible, both to be able to focus accurately and to make its illumination easier.

The Equipment

To photograph small objects you need a camera that will allow you to achieve sharp focus from a short distance, often as little as an inch or two. Specialized macro lenses are available for all SLR cameras, including recent digital models. They are relatively inexpensive, at least compared to the telephoto lenses used to photograph large wildlife. Many consumer digital cameras with a fixed lens can also be used to take spectacular macro shots, and their LCD screens help to frame and focus on your subject. The screen can often be swiveled or tilted, allowing you to take pictures from many different angles without having to contort your entire body in order to peek into the viewfinder.

I am frequently asked what particular camera models and lenses I use. New models come out every few months, but I feel that once you find a camera that fulfills your needs and are comfortable using it, there is no need to change it. If you already own an SLR and would like to venture into the field of macrophotography, invest in either inexpensive extension rings or a dedicated macro lens. But if you are new to photography, consider one of the many new models of digital cameras, paying close attention to its ability to focus on close objects rather than the number of megapixels it is boasting.

To take the images appearing in this book I employed a wide range of equipment, both film-based and digital. Many were taken with Nikon 90s and Nikon D1x, but more recently I started using Canon D1 Mark II series and Canon 10D. The lenses I find particularly useful for photographing smaller animals fall into the range of 60–180 mm macro lenses, although I am particularly fond of wide-angle 16–35 mm lenses, which equipped with a short extension ring provide a unique, open perspective rarely seen in macrophotography.

What should you photograph? This is entirely up to you. As long as you keep your mind open and your

eyes close to the ground and vegetation, you are bound to find fascinating subjects for your photography. But it is extremely helpful to know a few facts about your intended subjects, such as their preferred host plants, time of courtship activity, or a favorite habitat. Few of my photographs are a result of chance encounters. I usually know what animal or behavior I would like to capture with my camera and set out into the forest knowing what I am looking for. Do not be afraid to experiment with unusual angles or lighting, but avoid the two pitfalls of insect macrophotography. The first one is the temptation to chill or even kill these animals to avoid having to deal with their movement. Photographs taken this way are not only deceitful but also easily recognized as such by an experienced photographer or biologist. The second one concerns the way some animal photographs are displayed or printed. Many insects and other small animals prefer to rest upside down, but some photographers and editors find it disconcerting and "correct" them by rotating them 180°. I have seen some otherwise great photographs of insects published this way, some even showing insects in the process of molting but facing up in defiance of the laws of gravity.

In the end, no matter what you choose to focus on, simply lowering your gaze and getting a little closer to your subject can become one of your most memorable photographic experiences. Just keep your eyes wide open, walk slowly, look under your feet, and keep your camera gear dry. After a while you may discover that you too love the smaller majority.

Resources

Protecting endangered insects, mollusks, frogs, and other members of the smaller majority is the goal of several nonprofit organizations. If you would like to learn more about their work or make a donation to promote conservation of some of the most important members of the world's biosphere, please contact one of the following organizations.

Invertebrate Diversity Initiative
    (CABS, Conservation International)
    Museum of Comparative Zoology,
    Harvard University
    26 Oxford St., Cambridge, MA 02138, USA
    telephone: 617.496.8179
    http://www.smallermajority.org

Conservation International
    1616 M St. NW, Suite 600,
    Washington, DC 20036, USA
    telephone: 202.912.1000
    http://www.conservation.org

The Xerces Society for Invertebrate Conservation
    4828 SE Hawthorne Blvd.,
    Portland, OR 97215, USA
    telephone: 503.232.6639; fax: 503.233.6794
    e-mail: info@xerces.org
    http://www.xerces.org

IUCN (International Union for Conservation
    of Nature and Natural Resources)
    IUCN Species Survival Commission UK Office
    219c Huntingdon Rd.,
    Cambridge CB3 0DL, UK
    telephone: +44 (0)1223 277966
    fax +44 (0)1223 277845
    e-mail: redlist@ssc-uk.org;
    http://www.redlist.org

Nature Conservancy
    4245 North Fairfax Dr., Suite 100,
    Arlington, VA 22203, USA
    http://nature.org

NatureServe
    1101 Wilson Blvd., 15th floor,
    Arlington, VA 22209, USA
    telephone: 703.908.1800; fax 703.908.1917
    http://natureserve.org

World Wildlife Fund
    1250 24th St. NW,
    Washington, DC 20037, USA
    telephone: 202.293.4800
    http://www.worldwildlife.org

To learn more about insects, frogs, and other fascinating organisms; their biology and behavior; their role in the functioning of our planet; and other issues faced by the smaller majority, I recommend the following publications.

Insects and other land invertebrates

Eisner, T. 2003. *For Love of Insects.* Cambridge, Mass.: Belknap Press, 464 pp.

Hölldobler, R. and E. O. Wilson. 1995. *Journey to the Ants: A Story of Scientific Exploration.* Cambridge, Mass.: Belknap Press, 240 pp.

Preston-Mafham, R. and K. Preston-Mafham. 1993. *Encyclopedia of Land Invertebrate Behavior.* Cambridge, Mass.: The MIT Press, 320 pp.

Rentz, D. C. F. 1996. *Grasshopper Country: The Abundant Orthopteroid Insects of Australia.* Sydney: University of New South Wales Press, 284 pp.

Amphibians and reptiles

Pianka, E. R. and L. J. Vitt. 2003. *Lizards: Windows to the Evolution of Diversity.* Berkeley: University of California Press, 348 pp.

Savage, J. M. 2002. *The Amphibians and Reptiles of Costa Rica: A Herpetofauna between Two Continents, between Two Seas.* Chicago: University of Chicago Press, 954 pp.

Biodiversity conservation

Beattie, A. and P. R. Ehrlich. 2001. *Wild Solutions.* New Haven: Yale University Press, 256 pp.

Wilson, E. O., ed. 1988. *Biodiversity.* National Academy Press, 521 pp.

Acknowledgments

Throughout my life I have been extremely fortunate to meet and work with a number of fantastic people who made my dreams of exploration of the world's smaller majority possible. First and foremost I would like to thank my colleagues at Conservation International for their support and invitation to participate in their important work to save the world's diversity of life. Leeanne Alonso and Jennifer McCullough have invited me to join several of the Rapid Assessment Program expeditions, while Leslie Rice has always managed to keep me out of trouble, miraculously solving problems I tend to get myself into while on the road. A large portion of the images in this book have been made possible thanks to my participation in Project ALAS in Costa Rica, and I would like to thank Jack Longino and Robert K. Colwell for making me a part of it. I am indebted to the parataxonomists of ALAS, Maylin Paniagua, Danilo Brenes, Ronald Vargas, and Flor Cascante whose help has always been indispensable. Special thanks are owed to Kenji Nishida, a frequent travel companion in Costa Rica and an inspiration behind some of my photographs.

I am grateful to David C. F. Rentz, Brian Farrell, Brian Fisher, Daniel Otte, and Derek Sikes for either inviting me on or helping organize some of the trips during which I took images appearing in this book. Derek's comments and criticism were invaluable during the preparation of the manuscript. Other people whose comments on the manuscript have been very helpful include Leeanne Alonso, Gary Alpert, Jack Longino, Mark-Oliver Rödel, Simon Stuart, and David Wagner. I am greatly indebted to Edward O. Wilson for reviewing parts of the manuscript as well as for his help and encouragement during the initial stages of preparation of this publication. His philosophy and writing have been a major formative force in my development as a biologist and human being.

Many biologists have helped me identify species appearing on the pages of this book. For this I am grateful to Rogerio Bertani, Paul Brock, Stefan Cover, Neil Cumberlidge, Lee Dayer, Fabian Haas, Jonathan Leeming, Jack Longino, Wayne Maddison, William Muchmore, members of the Odonata mailing list, Jerry Powell, Lorenzo Prendini, Heather Proctor, Christopher J. Raxworthy, Mark-Oliver Rödel, Sacha Spector, David Wagner, Joseph Warfel, Rick West, Pawel Wieladek, Zhi-Qiang Zhang, and Manuel Zumbado.

My sincere thanks are due to the staff of Harvard University Press, most notably the book's designer Tim Jones, the editors Ann Downer-Hazell and Kate Brick, and the production coordinator David Foss.

Last but not least I will always be grateful to my wife, Kristin M. Smith, for her love and unwavering support.

Species Index